T0326387

Knowledge transfer in cattle husbandry

The EAAP series is published under the direction of Dr. P. Rafai

EAAP – European Association for Animal Production

The European Association for Animal Production wishes to express its appreciation to the *Ministero per le Politiche Agricole e Forestali* and the *Associazione Italiana Allevatori* for their valuable support of its activities

Knowledge transfer in cattle husbandry

New management practices, attitudes and adaptation

EAAP publication No. 117

Editors:

Abele Kuipers, Marija Klopcic, Cled Thomas

Wageningen Academic
P u b l i s h e r s

Subject headings:
Management practices
Attitudes
Adaptation

ISBN 9076998809
ISSN 0071-2477

First published, 2005

Wageningen Academic Publishers
The Netherlands, 2005

The designations employed and the presentation of material in this publication do not imply the expression of any opinion whatsoever on the part of the European Association for Animal Production concerning the legal status of any country, territory, city or area or of its authorities, or concerning the delimitation of its frontiers or boundaries.

Table of Contents

Preface

Transfer and utilization of knowledge is of great importance to justify the research efforts made by many institutes and organizations. In this context, it is not surprising that more attention for the demand side is advocated nowadays. However, in animal production meetings processes of knowledge utilization and transfer are certainly not common topics. The EAAP meetings held in 2003 in Rome, Italy and in 2004 in Bled, Slovenia were different. In both meetings a session was devoted to principles and examples of knowledge transfer and dissemination of research results. Also different organizational forms of extension and knowledge transfer have been described. The focus was especially on the adoption of new management practices in the field and the processes involved in this. One important factor is the attitude of the potential user towards the knowledge product and more in general towards innovation.

Effective knowledge utilization requires that insights and applications developed in the animal sciences are combined with social science approaches. A relevant insight from innovation studies, for example, is that innovations consist of three dimensions:
- Hardware: the technical devices and zoological know-how available
- Software: the intentions of the participants to work together and have the same goals in mind
- Orgware: the organisational structure in which activities are embedded

Successful innovation depends on the balancing and integration of these three dimensions. Often, however, we see that some dimensions are overlooked. In the world of animal sciences the emphasis tends to be on the "hardware", because this is seen as the core business of knowledge organisations. The "orgware" and "software" dimensions are too often neglected as critical factors to success.

In this book, the "orgware" and "software" concepts also receive some attention. Organisational aspects cannot be ignored, because the results of research must be often implemented in a complex structure (chain) of producer and consumer oriented organisations. Software aspects as attitude, personal preference, view towards the future and having a successor or not determine significantly the likelihood of the adoption of an innovation.

The contributions in this book are partly derived from the presentations in the sessions mentioned and partly collected elsewhere. Some more theoretical papers are followed by practical examples of implementation in the field and descriptions of extension networks and services. A large variety of papers is presented: from the power of the researcher and consultant to the power of data banks to the power of the client and producer in the dissemination process; from input driven to demand driven.

I like to thank my co-editors, Marija Klopcic for the endless hours she spent in preparing the outlay of this book so carefully and Cled Thomas for providing the final touch.

We believe this book is of interest to all of you who deal in some way with knowledge exchange and transfer activities.

Abele Kuipers
Secretary of Cattle Commission of EAAP
Director Expertisecentre for Farm Management and Knowledge Transfer

From agricultural extension to communication for innovation[*]

Cees Leeuwis

Communication and Innovation Studies, Wageningen University, Hollandseweg 1, 6706 KN, Wageningen, The Netherlands

Summary

Building on the tradition of agricultural extension, this chapter discusses several changes in thinking regarding the relations between communication and innovation. Starting from the idea that network building, social learning and conflict management are key processes to be supported by communication professionals, critical observations are made regarding dominant forms of participatory practice, innovation policy and privatisation of research and extension. Subsequently, the chapter discusses the implications of the proposed modes of thinking for the role of scientists in innovation processes. It is argued that a key role of scientists is to explicate implicit assumptions, claims and knowledge gaps in social learning processes, and to engage in collaborative research with societal stakeholders on a coherent set of natural and social science questions. The proposed conceptual models and observations eventually lead to the formulation of a research agenda for the field of Communication and Innovation Studies.

Keywords: communication, innovation, observations, role of scientists

Introduction

At Wageningen University, the field of Extension Education has been renamed into Communication and Innovation Studies. This change of title indicates the shift of focus from 'advisory communication' to the role of communication in more general terms during processes of change. At the root of this are all kinds of changes in conceptual thinking, some of which are linked to huge changes in context. Starting with the latter, it is clear that - in the Netherlands and elsewhere - the role of agriculture in rural society has become an increasingly controversial issue (Marsden, 1995). More and more interest groups have entered the debates about the use of land and rural space. Where increasing productivity and income in agriculture were once the central concerns, we now increasingly discuss matters like multifunctional land use, ecological services, food safety and the management of production chains. Difficult balances need to be struck constantly between competing goals and values in society. Against this background - and also due to developments within the discipline itself - our thinking about the two main concepts alluded to in the title of the present group - communication and innovation - has changed significantly.

---

[*] This chapter is a slightly modified and retitled version of an article that appeared in the Journal of Agricultural Education and Extension (2004, Vol 10, Nr.3). It orginates from an inaugural speech and is reprinted here with permission of the Journal.

Evolving thinking about communication

During the early years of Extension Education we had a rather mechanical and isolated idea of communication processes. At the time we thought in terms of individual senders and receivers who exchanged messages (i.e. transferred knowledge) through channels and media. If a message did not get across or was not understood, then 'interference' or 'static' was taken to be the cause. It soon became clear, however, that things were more complicated than this. Especially the idea that a message had a fixed meaning was found to be debatable. The receivers' interpretation of a message was usually quite different from that of the sender. This was caused not so much by 'noise' as by the fact that the sender and the receivers had very different frames of reference and prior knowledge. If a sender wanted to get a certain message across, it was concluded, they had to enter into the lifeworld of the receiver and had to be prepared to listen as well as send (Van den Ban, 1974; Dervin, 1981; Röling & Engel, 1990; Bosman *et al.*, 1989). Although this way of thinking was a big improvement, there were still some shortcomings. It frequently happened that senders were doing their very best to anticipate the receivers' lifeworld, only to see them almost refuse to understand the message. Despite all kinds of communicative efforts by government and industry, for example, Dutch farmers still have a rather negative perception about policies aimed at e.g. reducing mineral application or stimulating nature development, while consumers are not very willing to adapt their views regarding biotechnology. In these cases, it is not so much a question of incomprehension or lack of effective communication, but rather of an active and more or less purposeful maintaining of a difference in perception. In light of these kinds of experiences we now regard communication as a phenomenon in which those involved construct meanings (Leeuwis, 1993; Te Molder, 1995). Differences in interpretation have to do not only with different prior knowledge but also with other contextual issues such as the historically grown relationship between the communicating parties, configurations of interests, and also the influence of other actors not directly involved in the interaction. The negative reception of information regarding the governments' mineral management policy can only be understood against the background of the contentious history of the policy, the constant changes in policy, the contradictions between the mineral management policy and other policies, the economic interests of the farmers, the huge administrative burden on farm enterprises, and the polarization and enormous mutual distrust between farmers and government institutions. Also, the farmers' perceptions are influenced and shaped by other actors such as the feed industry, the fertilizer companies, accountants and the general public. In short, meanings come about - are actively constructed - in a complex context, and are not neutral. And communication is not something that necessarily brings people closer together or aids in problem-solving, but it can also add to incomprehension and the creation and up-keep of problems and conflicts (Aarts, 1998). So far some of the changes in thinking about communication.

Evolving thinking about innovation and change

Over the years, ideas about innovation and change have also evolved considerably. The fact that the advice given by agricultural extension workers, for example, was often ignored by the target groups eventually led to a reassessment of the quality of what was being proposed. The original hypothesis that innovations are developed by scientists, disseminated through extension and education and then put into practice by farmers and the public is called the linear innovation model, and has been refuted by many (Kline & Rosenberg, 1986; Röling, 1994; Rip, 1995). When one analyses successful innovation processes in retrospect, it is apparent that many ideas originate from practical experience and that the role of science is often limited[1].

Not only have the ideas about the origin of innovation changed, but also the ideas about what an innovation actually *is* are susceptible to transformation. During the early years of Extension Education, an innovation was regarded as something simple: a new type of plough or a new food product, for example. Moreover, the idea was that an innovation was either adopted or rejected by an individual, depending on all kinds of social conditions, among other things (Rogers, 1962; Van den Ban, 1974). It was thought that a new crop variety, for instance, could only be successful on the condition that certain input and output markets were adequately organised. Nowadays, we look at innovation differently. In the first place we recognise that innovations - even when considered solely from a technical perspective - are not one-dimensional[2], but must be viewed as large collections of partial innovations. Secondly, we do no longer regard the social and organisational conditions as external and static, but rather as integral parts of any innovation.

A illuminating example can be seen in Frank Van Schoubroeck's (1999) doctoral dissertation, written under supervision of Niels Röling and Joop Van Lenteren in 1999. Van Schoubroeck had been sent to Bhutan to develop a crop protection strategy to combat cutworm in maize, but eventually developed an integrated method to protect mandarin orchards from the Citrus fly. Through a long series of intermediate stages and dead-ends, an effective strategy was finally developed after many experiments in conjunction with the farmers in their orchards combined with research carried out at an experimental station. The strategy was based on the use of relatively small amounts of toxic bait, but also made big demands on the social organisation of the village. It was crucial that all farmers used the bait at the same point in time, in accordance with a specific developmental stage of the Citrus fly. In order to establish the exact moment, relatively intensive monitoring was necessary in the orchards. The community where the technology was developed managed to organise itself effectively: someone was put in charge of the monitoring, the poison and bait were bought and prepared communally and there was co-operation to ensure that all the orchards were treated at the same time. In a neighbouring community, however, this socio-technical solution did not work because a spiritual leader forbade the organised killing of insects (Van Schoubroeck, 1999).

Innovations do not just consist of new technical arrangements, therefore, but also of new social and organisational arrangements, such as new rules, perceptions, agreements and social relationships. This means, of course, that there are always many different stakeholders involved. It is a collective phenomenon in which social dilemmas (Koelen & Röling, 1994) and tensions are always likely to come to the fore. This means that it is not very useful to look at 'adoption' as something that happens at an individual level (as we thought in the past). What is important are the co-ordination and interdependencies between people. In line with Dirk Roep's (2000) dissertation, therefore, I would like to define innovation as 'a new pattern of co-ordination between people, technical devices and natural phenomena' (see also Smits, 2000 for a similar definition).

Finally, the thinking about innovation as a *process* has also changed dramatically over the past decades. In former days there was a strong belief in the possibility of planning and predicting change and innovation. In contrast, we now see that change is affected by complex inter-dependencies, fundamental uncertainties, chaos, unintended consequences, conflicts and unpredictable interactions that cannot be understood from a reductionistic perspective (Prigogine & Stengers, 1990; Holling, 1995). In connection with this, innovation processes are looked at nowadays from an evolutionary perspective. The idea is essentially that a variety of innovations and innovation processes compete in a dynamic selection environment in which the 'best fitting' survives (Bijker *et al.,* 1987; Rotmans *et al.,* 2001)[3]. What can be learned from this, among other things, is that sufficient variety must be created if one wishes to solve problems; it is important to back a number of horses (Van Woerkum & Aarts, 2002).

Against the background of these conceptual transformations there have, naturally, also been radical changes in our ideas about the links *between* communication and innovation. The focus has shifted from using communication as a means to transfer and effectuate knowledge, innovations and policies developed from the top down, to the study and organisation of communication and interaction in order to arrive at common starting-points, fitting and acceptable innovations and cogent policies. Thus, our ideas about the role of communication have undergone a 180-degree change in direction. Participation thus became an ever more important subject in research and in practice (Röling, 1996; Röling & Wagemakers, 1998; Van Woerkum, 1997).

A closer look at processes of communication and innovation

If we understand innovation as 'a new pattern of co-ordination between people, technical devices and natural phenomena', how should we understand innovation processes and what role does communication therein? Broadly speaking, there are two ways of dealing with this question. The first is descriptive/analytical: what is the actual role of communication? The second is a more normative approach: what could or should the role of communication be? Both starting points are of great interest to our discipline. One can only arrive at useful practical insights on the basis of a solid analytical understanding. Starting from a more normative viewpoint, I would like to suggest - building on the work of many members of the chairgroup - that three (simultaneous) processes deserve particular attention and communicative support.

Network building

The first process is that of the building of *networks*. Innovation requires co-ordinated action within a network of people. Such a network does not just spring into existence; it needs to be 'constructed'. And because renewal and innovation are at issue here, it will be evident that there is often a need for the forging of new relationships, both in terms of the parties involved and in terms of content (Engel, 1995), and for using these to expand windows of opportunity[4]. This may sound simple, but it is often not at all easy because, for instance, existing networks tend to close their doors to 'outsiders', or because certain parties just do not feel that they can be of any use to one another.

Social learning

At the same time that the building of a network is taking place, something that can be described as a *social learning process* must also occur. This means that the parties involved slowly develop overlapping - or at least complementary - goals, insights, interests and starting-points (Röling, 2002), and also build mutual trust and feelings of dependence and responsibility.

This is not 'learning' in the sense of 'knowledge transfer' or 'teaching'; rather it is about the development of different perspectives on reality through interaction with others. It is not just a question of cognitions about the natural and physical world but also of perceptions regarding one's own aspirations, abilities, responsibilities and space for maneouvre, and of other people's views of reality (see Figure 1). Exploration of different perspectives is vital in such a learning process because it is a very important route to 'reframing' (Gray, 1997): learning to look at a situation and one's role in it in a different way.

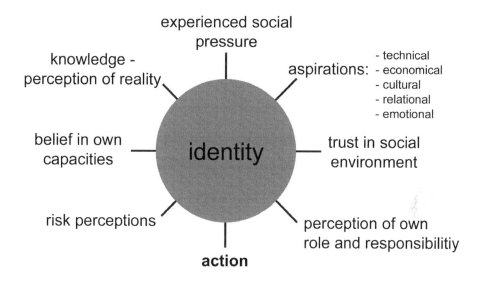

Figure 1. Different areas of perception (reflecting simultaneously reasons for action) that may be subject to 'learning' i.e. perceptual change (adapted and expanded from Röling, 2002).

Negotiation

A third process is that of negotiation and conflict management. Innovation implies changes in the status quo, which is always accompanied by friction and tension, especially in the case of innovations that go further than just optimisation within established frameworks and goals. Such innovations, which are characterised by the letting-go of existing starting points, goals and assumptions are also known as 'system innovations' or 'transitions' (Rotmans *et al.*, 2001; Geels, 2002). This kind of innovation and change brings with it, by definition, conflicts of interest between the parties involved and also with the established social and technological system or 'regime' that in many ways needs to be 'conquered' (Rip, 1995). In order to deal with such tensions, and in order to make new agreements and social arrangements, negotiation is essential. Preferably integrative negotiation[5] based on a social learning process (Aarts, 1998).

In view of the above, these three processes should guide and direct communicative intervention aimed at supporting innovation. This means that communications experts must lend their support to a large number of tasks that can be derived from theories about network building, social learning and negotiation. Tasks that are of great importance from the point of view of social learning might be: making the invisible visible, organising comparisons between different contexts, setting up experiments and facilitating exploration. A variety of communicative methods exist to support all this, ranging from dialogue and discussion techniques to model-based explorations (see Leeuwis with Van den Ban, in preparation). In addition, negotiation literature emphasises tasks such as the making and keeping procedural agreements, joint research and uncertainty reduction, guiding the give-and-take process, communication with constituencies and monitoring the observance of any agreements reached (Van Meegeren & Leeuwis, 1999). I deliberately do not use the word 'phases' here because we are dealing with issues that remain topical throughout the lifetime of the process. Moreover, in the context of innovation, learning processes happen on a variety of fronts, and and negotiation takes place with regard to a range of issues, and at different social levels. In short, we are dealing with complex and capricious series of interrelated events, with inherently unpredictable dynamics and results, the course of which can never be planned or controlled by a communications expert or facilitator[6]. Communications experts can, however, monitor the process and can facilitate progress at certain points.

It is interesting to note that, based on theories about networks building, social learning and negotiation, one can deduce a number of conditions and circumstances which affect the probability of achieving a productive innovation process. Learning, for example, costs time and energy and often fosters uncertainty (Aarts & Van Woerkum, 2002). It is, therefore, something that people only tend to do under certain circumstances. In order to engage in learning, people need to experience a serious problem, for example, preferably one that is urgent and visible. It is also important that people are confident that their learning will bear fruit, and that their social environment welcomes and gives space to a different perspective (Martijn & Koelen, 1999; Leeuwis, 2002). Negotiation literature also provides important pre-conditions, such as the insight that productive negotiation is only possible between parties who feel dependent on one another for solving a problematic situation (Aarts, 1998), which implies that there exists a certain balance of power, among other things. In other words, it is not possible to simply start or create an interactive innovation process from any situation.

Some observations regarding current practice

When we look at the day-to-day practice of innovation and communication from the above formulated perspective, what do we see? What is most striking? The following observations are limited mainly to communication and innovation in so far as they concern publicly formulated goals, such as sustainability, development, management of natural resources, health promotion, etc.

The need to reflect on pre-conditions

As has been touched upon in the preceding section, the stimulation of interactive innovation processes is not always useful or likely to be succesful, as certain pre-conditions may not be met. In practice we see that there is not much heed being paid to this, however. At times one gets the impression that everything nowadays must happen in as participatory and interactive a fashion as possible. Regardles of any pre-conditions, we see that every effort is made to bring all the different stakeholders together around the table or 'under the tree' in order for them to undergo a learning process together. In some cases participative trajectories have even become a fixed part of a bureaucratic formula. In Africa, for example, almost every rural project must begin with a standard selection of procedures, such as doing a 'transect walk', drawing up a 'seasonal calendar', and engaging in various 'ranking' exercises. Project workers do it because it has to be done and the communities have become used to it, often having results ready more or less off the shelf. In short, all kinds of rituals lacking any real content, influence or critical dialogue are frequently adhered to (Eyben & Ladbury, 1995; Graig & Porter, 1997; Pijnenburg, in preparation). But even when the intentions of the intervening parties are more carefully considered, we can see that the quality of social learning and negotiation processes often leaves much to be desired. We frequently witness, for example, that in-depth exploration is often a non-starter, that conflicts and issues of power are not brought to light, that results end up in the metaphorical 'bottom drawer', or that unattractive psuedo-compromises are reached which are then ignored by those involved (Leeuwis, 2000). All this is caused not only by a lack of ability on the part of the process facilitators but also by the fact that vital pre-conditions have not been met. There is often a lack of real institutional and political manoeuvring space, certain crucial stakeholders do not experience a problem, or the different stakeholders do not feel dependent on one another (Aarts, 1998). Under these kinds of conditions, the time is not (yet) ripe for an interactive innovation process and it would be much more useful - from the point of view of intervention - to try to create better conditions for such a process with the aid of more traditional (policy) instruments (Van Woerkum, 1990).With the aid of rule-making, political pressure, financial incentives and persuasive campaigns, for example, a feeling of interdependence can be fostered among the stakeholders in a river-basin, creating thereby a basis for a serious discussion about water distribution and utilisation. All this implies that it is time to move on from the idea that 'top-down' intervention and an interactive approach are mutually exclusive routes towards change and innovation; a concept which has become dominant in professional practice. Before, during and after an interactive innovation trajectories, strategic interventions can be extremely useful, while leadership and the taking on of responsibility are indispensable.

Knowledge policy or innovation policy?

Even though the ideas about innovation and the role of knowledge therein have changed dramatically, many researchers and advisors all over the world are still working according to an outdated task model. Attention is mostly focused on the technical side of the equation, i.e. developing and transferring technical knowledge and devices. Issues such as process management and the creation of coherence between social and technological arrangements receive relatively little attention. In many countries, the emphasis is still on knowledge transfer and knowledge policy rather than on innovation policy. In the Netherlands we witness a rather remarkable situation in this respect. The Ministry of Agriculture, Nature Conservation and Fisheries produced a document on innovation policy in 2001, containing many valuable ideas about innovation and change (Ministerie van LNV, 2001), but everyday policy practice within important policy domains seems to have little relation with this. On the subject of mineral management policy, for example, we can see that policy and project documents are full to bursting with 'knowledge development', 'knowledge transfer' and 'knowledge dissemination', while lack of technical expertise is actually of secondary importance in the case of this particular issue. It is much more a question of troubled relations, lack of agreement and a lack of coherence and co-ordination in the social network of which farmers are a part. These matters, however, do not really receive any explicit attention, partly because it does not fit into the mandate or role conception of the extension organisations and research institutes involved in the policy implementation. Somewhat similar phenomena can be witnessed with regard to (knowledge) policies in the sphere of organic agriculture[7]. In all, there seems to be a lack of congruence between innovation policy and knowledge policy.

Paradoxes of the 'knowledge market'

Despite the fact that role conceptions have not changed very much, there is visible movement in other areas of the knowledge infrastructure. All over the globe there are experiments being carried out with the privatisation of agricultural advisory services and with making research organisations independent from the state (Rivera & Zijp, 2002). One could say that a 'knowledge market' is being created. It's exact features differ from country to country. In most cases government agencies remain an important player on the market; thankfully governments still fund research and extension services on behalf of public issues. But an important change from the past is that research and extension organisations now have to compete in order to get access to these funds. The main intention in all this is that a 'knowledge market' will lead to a better balance between the 'supply' of and 'demand' for knowledge (Rivera & Gustafson, 1991; Umali & Schwarz, 1994). The term 'demand-driven service provision' has become a magic phrase. Looking at the effects of this situation, it is evident that there is a shift towards the servicing of well-funded customers and questions, especially in areas regarded by government as 'private' (Kidd *et al.,* 2000; Hanson & Just, 2001; Katz, 2002). When it comes to public issues, the role of governments - as financiers - in defining relevant questions is often quite big, which is somewhat strange in light of all the rhetoric about demand-orientation. But we will not go into that here.

What interests me most in view of the issue of innovation is the question of how new forms of funding influence the dynamics of innovation processes in the public interest. That is, processes of network building, learning and negotiation in precisely *those* situations where accepted solutions and ready-made knowledge products do not yet exist. If, for example, we examine the Dutch 'knowledge market' in the area of agriculture and rural resource management from this perspective, a number of remarkable features come to the fore:

Restrictions in knowledge exchange

Despite the fact that innovation processes benefit in theory from openness and the exchange of knowledge between the different parties, we have seen that co-operation between those involved in the knowledge network has become less self-evident. In the agricultural sector, the applied research and extension organisations have grown apart because of competition between them. There are also clear signs that farmers and horticulturists have become less willing to show their whole hand when taking part in study groups, partly because they have to pay more and more for knowledge (Oerlemans *et al.,* 1997). Lastly, the co-operation and exchange between the various publicly-funded projects - which are carried out by competing agencies - are often far from optimal. It should also be noted that these projects often have a limited life-span and have different starting and completion dates, making them difficult to link one to another. There is not much left of the famous - though by no means perfect - Dutch triptych of agricultural research, extension and education.

A lack of space for innovation

A second paradox is that - in order to be able to use a tendering system - the government must define (or have others define) fairly precisely what it wants in return for the money it provides for extension and research in the public interest. Concrete 'outputs' must be put down on paper. This provides a lot of clarity to project implementers with regard to the number of working days available and the activities to be carried out, but can be very limiting in innovation processes. One essential characteristic of learning and negotiation processes in multiple-stakeholder situations is that it is impossible to predict beforehand what the results will be and which directions for searching solutions will be agreed upon. Room to manoeuvre and flexibility are needed - not least in terms of the application of funds - but it has become apparent that this is not so easy to realise in the case of output-oriented financing[8]. Defining outputs beforehand can also lead easily to a certain level of inertia in areas for which outputs have not yet been defined. To put it bluntly, activities that are commonly understood as being relevant are left undone because nobody can justify and allocate the necessary man-hours in terms of the contracts available.

Transaction costs

In line with the above, a third interesting fact is that the stimulation of innovation processes through a 'knowledge market' and 'output-oriented funding' is often accompanied by large - sometimes vast - transaction costs. It is certainly not an easy task, and especially so for farmers and citizens, to get funding for a good innovative idea. One must ensure that the idea is defined somewhere as an output and then do everything in one's power in order to get an opportunity to tender for one's own idea. Millions of euros worth of man-hours are committed to presenting so-called 'business plans' to the Dutch government's innovation prgrammes and to those of the European Union. Project offices and consultants - referred to by farmers as the 'suit-and-tie culture' (Aarnink, pers. comm.) - are doing well out of it, but it has proven to be extremely difficult to link the course of these kinds of programmes (characterised by a long slow preamble, a few years in which everything must be completed, and then a return to inaction) to the dynamics of already existing innovation initiatives in society. The realisation that innovation processes are really all about new forms of co-ordination between different societal agents must lead us to conclude that innovation takes place primarily within society itself and not within the artificial boundaries of a project or programme. These boundaries can be very useful but must not be given a central role, though this is all too often what happens.

Conclusion

The preceding observations lead to the question of whether innovation in the public arena should really be supported in the form of a 'market'. The current arrangements are, in any case, attended by risks in the areas of co-operation, coherence, learning ability and pro-activeness. In my view, the problem starts already with the very idea of a 'supply' of and 'demand' for knowledge. These terms suggest immediately that there is somebody seeking knowledge and somebody supplying it, with the former having to pay the latter. This may perhaps work in the case of already tested and available advice and innovations, but not in a situation where new socio-technical innovations have to be developed[9]. In the latter contexts important questions tend to be unclear at the outset, while much of the relevant knowledge is still implicit, even if it is certain that many stakeholders have relevant knowledge to contribute. To put it another way, it is impossible to say who should pay whom and for what. In short, the idea of a 'knowledge market' is based on an overly simplistic and explicit conceptualisation of knowledge and also on a much too one-dimensional view: the 'market' is mainly oriented to substantive knowledge and not to other forms of perception and cognition that are important in an innovation context.

In the following section, I will set aside the practical obstacles signalled above and focus attention on the possible roles of scientists in innovation processes.

Knowledge and the role of the scientist in innovation

In innovation processes we are essentially faced with the challenge of linking all kinds of forms, domains, sources, and bearers of both knowledge and ignorance[10] to one another[11]. In connection with this it would be overly simplistic to consider 'knowledge' as being only a mental capacity. Knowledge and action are two sides of the same coin; a lot of knowledge seems to be 'stored' in our bodies and in the things around us, and is expressed through our actions, without our even consciously or actively reflecting on it (Giddens, 1984; Nonaka & Takeuchi, 1995; Scott, 1998). Knowledge is therefore often implicit; a large part of what we think, know, feel and are able to do is difficult to put into words. And even when we *are* able to put into words - i.e. if we communicate with others- we are usually more or less strategically selective in the words we use. Knowledge is, in short, an extremely elusive phenomenon. In light of this, how should we define the possible role of science? And what about the relationship between the natural and the social sciences?

Before addressing these questions, it is perhaps important to establish what we understand by the term 'science'. I would characterise scientific research as a subculture in which much importance is given to the development of original, valid and credible conclusions about reality[12]. Within the scientific community, there exist all kinds of epistemological subdivisions because there exist large differences between various groups of scientists regarding the way in which they arrive at their conclusions and the kinds of pronouncements that they make[13]. For this reason I prefer to use the phrase 'scientists' knowledge' rather than 'scientific knowledge'.

Role perception from an innovation perspective

Scientists in the domain of agriculture and natural resource management often have to deal with complex connections between technical, ecological, economic and social systems. There is much unpredictability and uncertainty and there are divergent values and interests at issue. This is precisely the kind of situation in which the philosophers Funtowicz & Ravetz (1993) argue for a post-normal approach to science, instead of a strategy in which science is only applied for the 'solving of puzzles' or the giving of situation-specific advice. With post-normal science, the scientists themselves are intensely involved in societal processes, discussions and innovation[14]. In other words, in processes of network building[15], social learning and negotiation.

In such contexts, the reaching of an agreement between the parties is often hampered by a lack of insight into certain issues or because there is a high level of uncertainty in technical and/or social areas. It is also possible that the available insights are not sufficiently explicit. All kinds of implicit claims to knowledge, assumptions and knowledge gaps are concealed in any communication between the parties. It can be important to make these explicit and open to discussion in order to assist the advance of an innovation process. This is not at all an easy task and will never be completely successful. Not only process facilitators but also scientists from various disciplines can play an important role in this respect. One may expect scientists to have a certain sensitivity regarding implicit assumptions, claims and knowledge gaps in their own areas of expertise. A serious dialogue between scientists and societal stakeholders, in which the different parties have the opportunity to ask each other difficult questions, can contribute to making explicit previously implicit issues. If knowledge gaps also arise during this dialogue then the presence of researchers will naturally be helpful in developing answers with the aid of research. From the point of view of negotiation, conducting joint research is what most relevant. That is; research in which various stakeholders are involved closely in the refinement of research questions, the choice of methods and the fixing of the research location (a laboratory, an experimental station, a computer model or a field situation). This is because it is important not only to generate answers, but also that the parties involved have confidence in the results. In addition, collaboration in carrying out research can contribute to an improvement in the relationship between the stakeholders involved (Van Meegeren & Leeuwis, 1999).

This does not imply, however, that nothing remains of the individual responsibility and autonomy of the researcher. Here it is relevant to note that a crucial trigger for social learning is feedback (Kolb, 1984; Heymann, 1999). In innovation processes, therefore, both natural and social scientists can stimulate learning processes by providing - more or less confrontational - feedback at their own discretion. They can provide not only insights based on research with reference to that specific situation but also those gleaned elsewhere, or they can make projections about the future or point to radically different technological or social solutions.

The status of knowledge contributed by scientists

Some natural (and also social) scientists may have winced when reading the above. Not so much because I attribute a somewhat modest role to scientific researchers in innovation processes - many natural scientists are far more modest about their role than at times portrayed by social scientists - but because I have given very little attention to the role of scientists as 'referees' in situations where conflicting views on reality are at issue. Is it not the task of science to bring the truth to light? In my experience, many natural scientists feel threatened by the idea that reality is something that is constructed. It could, after all, lead to a situation where the scientist's perspective is pushed aside as being just one of the many equally valid views on reality! This is not what I am advocating. It seems to me that it remains possible and important to differentiate between sense and nonsense, and between more and less well-founded views on reality[16]. In my opinion, the essence of constructivism is not so much that every truth is relative but rather that every truth has its limits and also that in everyday life neutral truths do not exist.

When, for example, a laboratory experiment shows a link to between the presence of the nitrogen fixating bacteria Rhizobium and grop growth, this can lead to a conclusion that is valid within the context of the experiment. That is to say: given a particular type of soil, particular climatological conditions, a particular labour input, a particular form of crop protection, a particular planting date, etc[17]. In other words, the conclusions drawn from the experiment are only valid within the limits of its context. Many of the conditions outside the laboratory and/or experimental station will most probably be quite different. When knowledge that is valid within a certain local context (the laboratory or experimental station in this example) is transplanted directly into a different local context (an agricultural region, for example) there are bound to be problems. To put it bluntly: scientific knowledge too is a form of local knowledge.

One important aspect of such local specificity is connected with my second point; namely, the fact that neutral truths do not exist. This has to do with the fact that a particular research initiative is usually brought about by a particular issue. The question of whether there is a link between the presence of Rhizobium and crop growth is not at all a neutral one, but arises from a certain problem perception and is therefore linked to social aims. It is not a question that is likely to be brought up by the fertilizer industry but it is likely to be asked by organic farmers and development organisations. And if questions are not neutral, then the answers will not be either. Answers are used by people as 'weapons' in a 'struggle' with other interests; so it matters for which questions scientists try to formulate answers, and for which not.

In a nutshell, scientists have to realise that their knowledge has a local character and is not neutral. In connection with this Alroe & Kristensen (2002) argue for a 'reflexively objective science' in which scientists not only realise this but make it explicit and transparent. In other words, scientists should be expected to open up the hidden dimensions of their own research questions and knowledge to discussion. Such transparency does not mean that scientists will become politicians. The opposite is true, in fact. When scientists are clear about underlying social values and goals it can only become more obvious that coflicts of interest cannot be settled by scientists and that it is up to societal stakeholders, authorities and politicians to judge the value of the different view points and to make decisions.

Working across disciplinary boundaries

The foregoing is also connected with the manner in which co-operation between social and natural scientists can take shape[18]. The essence here, according to me, is that natural and social scientists influence and refine one another's assumptions, research questions and action plans. In other words, it is about putting the most relevant non-neutral questions on the agenda. These can also be very 'fundamental' questions[19]. One example of such mutual influencing can be taken from the 'Convergence of Sciences' project that is being co-ordinated by the entomologist Arnold Van Huis, in which nine doctoral students are being guided by both natural and social scientists from Wageningen University and universities in Ghana and Benin.

In an initial investigative stage of this project, the researchers in Ghana came across a complex crop-rotation system in which farmers attributed soil-fertility enhancing properties to a certain variety of cassava. This was interesting, because it ran directly counter to the accepted theory that cassava actually exhausts the soil. Doctoral student Samuel Adjei Nsiah set out to examine this system in greater depth and, where possible, improve it. Spurred on by his interest in the social aspects of this innovation, he eventually discovered that the rotation system is mainly applied by the native population of the area and not by the migrants who come from the north of Ghana. The latter are aware of the system but usually cannot apply it because they own no land and the locals will only agree to short-term leasing contracts. The latter, then, is associated with specific attitudes to money, inflation, the land tenure system, mutual distrust and with the role played by the local authorities (Adjei Nsiah *et al.,* in preparation). This example illustrates once again that diversity within communities is an important subject (Van der Ploeg, 1994; Hebinck & Ruben, 1998). We can also see that - from the point of view of the migrants - there is little advantage to be gained if the natural scientists concentrate solely on the further development of the multi-year rotation system, at least as long as nothing changes regarding the issue of contracts between landowners and tenants. It would, perhaps, be more useful to search with the migrants for single-year intercropping systems that would have an immediate effect on soil fertility. Furthermore, based on the insights gained, social scientific research could be directed towards bringing about a better understanding of the dilemmas faced by the native population and the migrants with reference to land use and leasing contracts, and towards identifying and mobilising bringing actors and institutions that could help to break the deadlock.

Such fine-tuning of natural and social science research questions is far from standard practice. For a broader application, new organisational forms, methods and tools for 'beta/gamma science' (Röling, 2000) are essential. There is still scope for immense progress in this area.

By means of conclusion: issues for research and education

What is the significance of the foregoing to future research and education in Communication and Innovation Studies?

Research themes

Over the coming years, socio-technical innovation processes and the role of communication therein will be a prominent theme. It is a research orientation that is useful not only in the area of rural development and sustainability but also in that of, for example, health promotion and agricultural chain management. Five inter-connected research topics stand out:

(1) *The construction of (in)coherence*: We have seen that innovation is about the achieving of effective co-ordination in a network of people, technical devices and natural phenomena. All this in the form of a coherent whole of new technical and social-organisational arrangements. In order to arrive at this kind of co-ordination and coherence, it is essential that different forms, kinds and sources of cognition be connected together in a social learning and negotiation process. It is important for our field that we begin to better understand how such connections come into existence, or do not, and what the role of communication is in all this. One approach to researching this is to zoom in on the moments in innovation processes at which certain break-throughs, changes in direction and selections occur, respectively on the moments at which the process stagnates. Looking back on episodes in which - in retrospect - things 'went awry' is also of interest, naturally. On the basis of a better understanding of these kinds of situations, new focus points for process facilitation can also be formulated.

Within this theme the value and influence of different communicative tools can be studied in connection with the ways in which they are used. Interesting tools here include, for example, methods for explicating knowledge, methods for demand articulation, and explorative techniques, including also model-based explorations (Rossing *et al.* 1999).

(2) *Beta/gamma interaction*: A second theme, which can be viewed as a special subject within the first, is the interaction between natural and social scientists, and also the relationship between beta and gamma knowledge. We must not only practise beta/gamma co-operation but we must also study it! How do processes of beta/gamma interaction and communication evolve in an innovation context, and why do things happen in this way? To what effect? What is the value of specific communicative techniques? This dual interest can be somewhat of a dilemma. On the one hand, we are often expected to contribute to the shaping of beta/gamma interaction and interactive processes, and even to play an active facilitating role. On the other hand, our desire to engage in critical reflection, analysis and theory formulation regarding such processes, necessitates not only active involvement but also a certain amount of critical distance. In my opinion, we should certainly contribute to design of processes, but should leave the actual facilitation to others.

(3) *The influence of the 'knowledge market' and other institutions*: One of the main themes of this article has been that communication on behalf of innovation is influenced not only by the backgrounds of the societal stakeholders involved but also by the manner in which intervention is integrated into organisational, administrative and/or financial structures[20]. Such institutional influences, therefore, form a third research theme, within which I would like to emphasise the manner in which the rise of a 'knowledge market' influences the course

of innovation processes in the public domain. Comparisons with similar innovation processes in different institutional contexts could be an interesting research approach in this area.

(4) *Alternation between 'top-down' and 'bottom-up'*: A fourth research theme concerns the alternation between 'top-down' and 'bottom-up' moments and interventions in socio-technical transformation processes. How do these to kinds of strategies affect one another in practice? How does this influence different kinds of innovation processes in a positive or negative way? And what patterns of alternation are more and less useful? This theme is also closely linked to the question of the ways in which the processes of network building, social learning and negotiaton are (or can be) interwoven. Here too the issue of more and less productive forms of interaction is of interest.

(5) *A methodology for process monitoring*: Within the framework of methodology development, I would like to propose process monitoring as the fifth research topic. When we assume that processes are impossible to pre-plan in any useful manner, then this implies, among other things, that participants and process facilitators need to keep a constant eye on how processes evolve, and need to be able to respond to emergent developments. But because much of what is happening occurs, by definition, behind the scenes, the question of how process facilitators and participants can get a clearer insight into it is important. The development of a feasible approach and methodology for monitoring the course of learning and negotiation processes is therefore of interest.

For the study of the subjects mentioned above, it is important to develop approaches to research through which innovation processes - i.e. processes of network building, social learning and negotiation - can be followed and documented over time and compared with one another.

Developing a new kind of professional

In view of the developments in our thinking about communication, innovation and the relations between these phenomena, it is clear that we need to move beyond the classical 'extension professional', whose work was mainly in the area of knowledge transfer, persuasion, providing individual advice and supporting horizontal knowledge exchange. Although such areas of activity can still be relevant for purposes of stimulating innovation, they need to be complemented by the provision of other key communicative services which are not traditionally associated with 'extension'. Such services include the facilitation of network building, multi-stakeholder learning and conflict management which constitute key processes for arriving at new (and coherent) technical and social-organisational arrangements. Thus, we need to train bridge-builders who can mobilise and link the expertise of natural scientists, social scientists and societal stakeholders, and who can communicatively support complex multi-actor processes in order to enhance opportunities for arriving at new patterns of coordinated action. Clearly, this requires analytical and practical competencies that deviate considerably from those with whom classical 'extensionists' were equipped.

References

Aarts, M.N.C., 1998. Een kwestie van natuur; een studie naar de aard en het verloop van communicatie over natuur en natuurbeleid. Dissertation. Landbouwuniversiteit Wageningen, Wageningen, The Netherlands.

Aarts, M.N.C. & C.M.J. Van Woerkum, 2002. Dealing with uncertainty in solving complex problems. In: Leeuwis, C. & R. Pyburn (editors), 421-435.

Adjei Nsiah, S., K. Giller, T. Kuyper, W. Van der Werf, K. Abekoe, O. Sakyi-Dawson & C. Leeuwis (in preparation). Explaining the differential uptake of locally developed soil fertility management innovations in Ghana. Concept article.

Alroe, H.& E.S. Kristensen, 2002. Towards a systemic research methodology in agriculture: Rethinking the role of values in science. Agriculture and Human Values, 19: 3-23

Arce, A. & N. Long, 1987. The dynamics of knowledge interfaces between Mexican agricultural bureaucrats and peasants: A case study from Jalisco. Boletín de Estudios Latinoamericanos y del Caribe, 43: 5-30

Bijker, W., T. Hughes & T. Pinch (editor), 1987. The social construction of technological systems. New directions in the sociology and history of technology. MIT-Press, Cambridge MA.

Bosman, J., E, Hollander, P. Nelissen, K. Renckstorf, F. Wester & C. Van Woerkum, 1989. Het omgaan met kennis - en de vraag naar voorlichting: Een multidisciplinair theoretisch referentiekader voor empirisch onderzoek naar de vraag naar voorlichting. Katholieke Universiteit Nijmegen, Nijmegen.

Bouma, J., 1999. The role of research chains and user interaction in designing multi-functional agricultural production systems. In: C. Leeuwis (editor), 219-235.

Craig, D. & D. Porter, 1997. Framing participation: development projects, professionals and organisations. Development in Practice, 7: 229-236

De Grip, K., C. Leeuwis & L.W.A. Klerkx, 2003. Ervaringen met het Steunpunt Mineralen concept. Lessen over vraagsturing. Agro Management Tools, Wageningen, The Netherlands.

Dervin, B., 1981. Mass communication: Changing conceptions of the audience. In: R.E. Rice & W.J. Paisley (editors), Public communication campaigns., Sage Publications, Beverly Hills., 71-88.

Engel, P.G.H., 1995. Facilitating innovation. An action-oriented and participatory methodology to improve innovative social practice in agriculture. Doctoral dissertation. Wageningen Agricultural University, Wageningen, The Netherlands

Eyben, R & S. Ladbury, 1995. Popular participation in aid-assisted projects: Why more in theory than in practice. In: Power and participatory development. Theory and practice. N. Nelson & S. Wright, Intermediate Technology Publications, London, United Kingdom, 192-200.

Funtowicz, S.O. & J.R. Ravetz, 1993. Science for the post-normal age. Futures, 25: 739-755

Geels, F., 2002. Understanding the dynamics of technological transitions. A co-evolutionary and socio-technical analysis. Twente University Press, Enschede.

Giddens, A., 1984. The constitution of society: Outline of the theory of structuration. Polity Press, Cambridge.

Gray, B., 1997. Framing and re-framing of intractable environmental disputes. Prentice Hall, London.

Hanson, J.C. & R.E. Just, 2001. The potential for transition to paid extension: Some guiding economic principles. American Journal of Agricultural Economics, 83: 777-784

Hebinck, P. & R. Ruben, 1998. Rural households and livelihood strategies: straddling farm and nonfarm activities. In: Proceedings of the 15th International Symposium on Farming Systems Research and Extension 'Going Beyond the Farm Boundary', Pretoria, 29 November - 4 December 1998., 876-885.

Heymann, F., 1999. Interpersoonlijke communicatie. In: C.M.J. Van Woerkum & R.C.F. Van Meegeren (editors), 174-194.

Holling, C.S., 1995. What barriers? What bridges? In: Barriers and bridges to the renewal of ecosystems and institutions. Gunderson, L.H., C.S. Holling & S.S. Light (editors). Colombia Press, New York., USA, 3-37.

Hounkonnou, D., 2001. Listen to the Cradle. Building from Local Dynamics for African Renaissance. Case studies in rural areas in Benin, Burkina Faso and Ghana. Doctoral dissertation. Wageningen University, Wageningen, The Netherlands.

Jiggins, J. & D. Gibbon, 1998. What does interdisciplinary mean? Experiences from SLU. In: The challenges for extension education in a changing rural world, A. Markey, J. Phelan & S. Wilson (editors). Proceedings of the 13th European Seminar on Extension Education, August 31-September 6, 1997, Department of Agribusiness, Extension and Rural Development. UCD, Dublin, Ireland, 317-325.

Katz, E. (with contributions by A. Barandun), 2002. Innovative approaches to financing extension for agricultural and natural resource management – Conceptual considerations and analysis of experience. LBL, Swiss Center for Agricultural Extension, Lindau, Switzerland.

Kidd, A.D., J.P.A. Lamers, P.P. Ficarelli & V. Hoffmann, 2000. Privatising agricultural extension: Caveat Emptor. Journal of Rural Studies, 16: 95-102

Kline, S.J. & N. Rosenberg, 1986. An overview of innovation. In: The positive sum strategy: Harnessing technology for economic growth. R. Landau & N. Rosenberg (editors), National Academic Press, Washington, USA, 275-305.

Knorr-Cetina, K.D., 1992. The couch, the cathedral, and the laboratory: On the relationship between experiment and laboratory in science. In: Science as practice and culture. A. Pickering (editor), University of Chicago Press, Chicago, USA, 113-138.

Kolb, D.A., 1984. Experiential learning: Experience as the source of learning and development. Prentice-Hall, Englewood Cliffs.

Koelen, M. A. & N.G. Röling, 1994. Sociale dilemma's. In: Basisboek voorlichtingskunde. Röling, N.G., Kuiper, D. & Janmaat, R. (editors), Boom, Amsterdam / Meppel, The Netherlands, 58-74.

Kuhn, T.S., 1970. The structure of scientific revolutions. University of Chicago Press, Chicago.

Leeuwis, C., 1993. Of computers, myths and modelling: The social construction of diversity, knowledge, information and communication technologies in Dutch horticulture and agricultural extension. Wageningen Studies in Sociology, Nr. 36. Wageningen Agricultural University, Wageningen, The Netherlands.

Leeuwis, C. (editor), 1999. Integral design: innovation in agriculture and resource management. Mansholt Institute / Backhuys Publishers, Wageningen / Leiden, The Netherlands, 123-143.

Leeuwis, C., 2000. Re-conceptualizing participation for sustainable rural development. Towards a negotiation approach. Development and Change, 31: 931-959

Leeuwis, C. & R. Pyburn (editors), 2002. Wheelbarrows full of frogs. Social learning in rural resource management. Royal Van Gorcum, Assen.

Leeuwis, C., 2002. Making explicit the social dimensions of cognition. In: Leeuwis, C. & R. Pyburn (editors), 391-406.

Leeuwis, C. & A. Van den Ban, 2004. (in preparation). Communication for rural innovation. Rethinking agricultural extension. Blackwell Science, Oxford.

Marsden, T.K., 1995. Beyond agriculture: regulating the new rural spaces. Journal of Rural Studies, 14: 285-297

Martijn, C. & M.A. Koelen, 1999. Persuasieve communicatie. In: C.M.J. Van Woerkum & R.C.F. Van Meegeren (editors), 78-104.

Ministerie van LNV, 2001. Innovatie: sleutel tot verandering. LNV innovatiebeleid voor Voedsel en Groen. Ministerie van LNV, Den Haag, The Netherlands.

Nonaka I. & H. Takeuchi, 1995. The knowledge creating company: how Japanese companies create the dynamics of innovation. Oxford University Press, New York / Oxford, USA.

Oerlemans, N., J. Proost & J. Rauwhorst, 1997. Farmers' study groups in the Netherlands. In: Farmers' research in practice. Lessons from the field. L. Veldhuizen, A. Waters-Bayer & R. Ramirez (editors), Intermediate Technology Publications, London, United Kingdom, 263-277.

Prigogine, I. & I. Stengers, 1990. Orde uit chaos: een nieuwe dialoog tussen de mens en de natuur. Uitgeverij Bert Bakker, Amsterdam, The Netherlands.

Pruitt, D.G. & P.J. Carnevale, 1993. Negotiation in social conflict. Open University Press, Buckingham.

Pijnenburg, B., 2004. (in preparation). Keeping it vague. Discourses and practices of participation in rural Mozambique. Draft for doctoral dissertation. Wageningen University, Wageningen, The Netherlands.

Richards, P., 1994. Local knowledge formation and validation: the case of rice production in central Sierra Leone. In: Beyond farmer first. Rural people's knowledge, agricultural research and extension practice. Scoones, I & J. Thompson (editors), Intermediate Technology Publications. London, United Kingdom, 165-170.

Rip, A., 1995. Introduction of new technology: making use of recent insights from sociology and economics of technology. Technology Analysis & Strategic Management, 7: 417-431

Rivera, W.M. & D.J. Gustafson (editors), 1991. Agricultural extension: worldwide institutional evolution & forces for change. Elsevier Science Publishers, Amsterdam, The Netherlands.

Rivera, W.M. & W. Zijp (editors), 2002. Contracting for agricultural extension. International case studies and emerging practices. CABI Publishing, Wallingford.

Roep, D., 2000. Vernieuwend werken. Sporen van vermogen en onvermogen. Dissertation. Wageningen Universiteit, Wageningen, The Netherlands.

Rogers, E.M., 1962. Diffusion of innovations, 1st edition. Free Press, New York, USA

Röling, N.G. & P.G.H. Engel, 1990. IT from a knowledge systems perspective: Concepts and issues. Knowledge in Society: The international journal of knowledge transfer, 3: 6-18

Röling, N.G., 1994. Voorlichting en innovatie. In: Basisboek voorlichtingskunde. Röling, N.G., Kuiper, D. & Janmaat, R. (editors), Boom, Amsterdam / Meppel, The Netherlands, 275-294.

Röling, N.G., 1996. Towards an interactive agricultural science. European Journal of Agricultural Education and Extension, 2: 35-48

Röling & M.A.E. Wagemakers (editors), 1998. Facilitating sustainable agriculture. Participatory learning and adaptive management in times of environmental uncertainty. Cambridge University Press, Cambridge.

Röling, N.G., 2000. Gateway to the global garden: beta/gamma science for dealing with ecological rationality. Eight Annual Hopper Lecture, October 24, 2000. University of Guelph, Guelph, Canada.

Röling, N.G., 2002. Beyond the aggregation of individual preferences. Moving from multiple to distributed cognition in resource dilemmas. In: Leeuwis, C. & R. Pyburn (editors), 25-47.

Rossing, W.H.A., M.K. Van Ittersum, H.F.M. Ten Berge & C. Leeuwis, 1999. Designing land use options and policies. Fostering co-operation between Kasparov and Deep Blue? In: Leeuwis, C. (editor), 49-72.

Rotmans, J., R. Kemp & M.B.A. Van Asselt, 2001. More evolution than revolution: transition management in public policy. Foresight, 3: 15-31

Scott, J.C., 1998. Seeing like a state: how certain schemes to improve the human condition have failed. Yale University Press, New Haven / London, United Kingdom.

Smits, R., 2000. Innovatie in de universiteit. Acceptance speech. Universiteit Utrecht, Utrecht, The Netherlands.

Te Molder, H., 1995. Discourse of Dilemmas: An Analysis of Government Communicators' Talk. Doctoral dissertation. Wageningen Agricultural University, Wageningen, The Netherlands.

Umali, D.L. & L. Schwarz, 1994. Public and private agricultural extension: beyond traditional boundaries. World Bank Discussion Paper. World Bank, Washington, USA.

Van den Ban, A.W., 1974. Inleiding tot de voorlichtingskunde. Boom, Meppel / Amsterdam, The Netherlands.

Van der Ploeg, J.D., 1991. Landbouw als mensenwerk: Arbeid en technologie in de agrarische ontwikkeling. Coutinho, Muiderberg.

Van der Ploeg, J.D., 1994. Styles of farming: an introductory note on concepts and methodology. In: Born from within: practices and perspectives of endogenous development. J.D. Van der Ploeg & A. Long (editors), Royal Van Gorcum, Assen, 7-30.

Van der Ploeg, J.D., 1999. De virtuele boer. Koninklijke Van Gorcum, Assen.

Van Meegeren, R.C.F. & C. Leeuwis, 1999. Towards an interactive design methodology: guidelines for communication. In: C. Leeuwis (editor), 205-217.

Van Schoubroeck, F.H.J., 1999. Learning to fight a fly: developing citrus IPM in Bhutan. Doctoral dissertation. Wageningen University, Wageningen, The Netherlands.

Van Woerkum, C.M.J., 1990. Het instrumentele nut van voorlichting in beleidsprocessen. Massacommunicatie, 18: 263-278

Van Woerkum, C.M. J., 1997. Communicatie en interactieve beleidsvorming. Bohn Stafleu Van Loghum, Houten / Diegem.

Van Woerkum, C.M.J. & R.C.F. Van Meegeren (editors), 1999. Basisboek communicatie en verandering. Boom, Amsterdam / Meppel, The Netherlands.

Van Woerkum, C.M.J. & M.N.C Aarts, 2002. Wat maakt het verschil. Over de waarde van pluriformiteit in interactieve beleidsprocessen. Innovatie Netwerk Groene Ruimte en Agrocluster, Den Haag, The Netherlands.

Winograd, T & C.F. Flores, 1986. Understanding computers and cognition: A new foundation for design. Ablex Publishing Corporation, Norwood.

NOTES

[1] Successful innovations appeared to be based on the effective integration of the problem perceptions, knowledge and experience of scientists, clients, intermediaries and other parties involved. Common starting points and goals regarding desired outcomes were also found to be important.

[2] If a new type of crop is introduced on a farm, then a lot of other things are likely to change as well; crop protection, animal diet, crop rotation and also social issues such as labour organisation, the relationship with buyers, etc. (Van der Ploeg, 1991).

[3] The concept of innovation proposed here implies that the selection environment is not only dynamic but can also be actively influenced, i.e.can be made an integral part of the innovation. For instance, the dominant selection environment in conventional modern agriculture (in which one could include the agro-industrial sector) is not a very fertile ground for organic agriculture, which is why people with a passion for organic agriculture have, in a manner of speaking, created their own selection environment with the aid of all kinds of parallel institutions and organisational structures.

[4] Multifunctional land use, for example, is only possible if new networks are formed between, farmers, the general public, consumers, water-use control bodies, etc.

[5] A relevant distinction in connection with this is that between distributive and integrative negotiation (Aarts, 1998; Pruitt & Carnevale, 1993). In distributive negotiations, the parties hold on to their existing perspectives and positions and the negotiations are mainly used to 'divide the cake'. Aarts (1998) points out that 'distributive compromises' tend not to be stable because the nature and form of the conflicts does not change. 'Integrative' negotiation processes, however, involve joint learning in which new problem definitions, perceptions and creative solutions are developed (see Aarts, 1998).

[6] It is clear from the foregoing that innovation is not really about planning or decision-making in the traditional sense. Neither is it about the communicative support of stages or tasks in a rational planning process. If only the classic project cycle is followed then fundamental issues will be overlooked.

[7] The majority of the monies made available for research purposes are spent on the development of technical knowledge, while the sector's main problems lie in other areas, for the most part.

[8] In reality, in the guise of 'demand orientation', we have seen the return of a top-down direction and planning of publicly financed extension and research. The principles of supply and demand are at work, but the government formulates the demand. This would not be so bad if the government - as the protector of the public interest - had a good sense of what is happening in the everyday practice of rural renewal, but - as Van der Ploeg (1999) points out - with the privatisation of advisory services, the government has lost its eyes and ears at the grass-roots level.

[9] Neither does it work in the case of advice given in connection with restrictive policies, which are quite common in the public arena. For example, with regard to the mineral management policy, there is a lack of active demand for knowledge on the part of farmers, which forms a serious obstacle to the achievement of a 'demand-oriented' market (Katz, 2002; De Grip et al., 2003).

[10] The phenomenon of 'knowledge' is inherently connected to ignorance (Winograd & Flores, 1986). With the aid of knowledge we not only lessen our ignorance but can also increase it. Not just because new knowledge leads to new questions but also because what we think we know for certain is a barrier to seeing other possible explanations and perspectives regarding reality (Arce & Long, 1987).

[11] Here we are dealing with, among other things, the many perceptions, valuations and practices of different social agents, which - to further complicate matters - operate at different societal and governmental aggregation levels. Similarly, it involves insights concerning different levels of scale (molecule, cell, organism, crop, ecosystem, etc.), different domains of human practice and disciplines.

[12] It is evident that most scientists use activities, methods and standards very different from those used by non-scientists to arrive at credible and valid conclusions. There are big differences, for example, between farmer experiments and those carried out by scientists (Richards, 1994; Leeuwis with Van den Ban, in preparation). Nevertheless, both can be useful, and both are associated with specific strong and weak points (Van Schoubroeck, 1999).

[13] The differences involved are not only those between the social and natural sciences; there is also diversity within these categories (Knorr-Cetina, 1992).

[14] In Wageningen University the term 'interactive science' is usually used in this context. This phrase does, perhaps, still place 'research' and 'science' rather too centrally in contexts of societal problem-solving and innovation.

[15] From the point of view of 'network building' it is important that scientists become part of the networks in which innovation is brought about. In other words, they must allow themselves to get involved in initiatives for change and innovation in society. There is usually no lack of such initiatives; however, they may not always be immediately visible, they may well exist in unlikely places (Hounkonnou, 2001), and the people involved are not always likely to come into contact with researchers. Thus, researchers have to actively search for such initiatives.

[16] In certain contexts, scientists can definitely make very well-founded and difficult-to-refute statements about a particular reality. A good example is a series of laboratory experiments, set up according to the classical reductionistic philosophy, and aimed at discovering certain causal relationships, e.g. those between the presence of Rhizobium bacteria and crop growth. In other cases, relating, for example, to complex ecosystems, the weather or global food production, it is far more difficult to arrive at unequivocal conclusions (Holling, 1995), but even here a distinction can probably still be made between more and less well-founded statements.

[17] And as long as underlying theories, concepts, measurement techniques, etc., remain undisputed (Kuhn, 1970).

[18] There are different views of what working across disciplinary boundaries entails, varying from simply 'adding up' isolated contributions from different disciplines, to the development of a special trans-disciplinary language within a multi-disciplinary team. Jiggins & Gibbon (1998) provide an interesting overview in this respect. In my view, the minimum requirement is that different scientists interact with each other in an effort to deal with a problematic situation.

[19] It would be a mistake to think that there is no room in interactive innovation processes for answering 'fundamental' questions (Bouma, 1999). It is probable, however, that the fundamental questions arising during an interactive process would be different from those arising during a discussion among only scientists.

[20] I hasten to point out that such structural influences do not operate behind the back of people; in their interaction human agents actively refer to - and make use of - structural properties (Giddens, 1984).

Different approaches to advise cattle farmers: the charter of good practices in cattle breeding

Anne-Charlotte Dockès and Caroline Hédouin

French Livestock Institute (Institut de l'Elevage), Project Engineering Group (Service Ingénierie de Projet), 149 rue de Bercy, 75595 Paris Cedex 12, France

Summary

Cattle farmers are permanently making technical and economic decisions about their farms. They use their knowledge, their social representations, discussions in social groups and the practical aspects involved. Advice can help them to make these decisions. Advisory operations can be managed as projects and take into account the diversity of farmers' expectations. A collective approach of advisory operations aims at maintaining the efficiency of the small group advice while applying to the greatest number of farmers. In this paper, we are going to present a Charter of 35 good practices which has been proposed to cattle farmers for 3 years. This charter concerns different aspects of the cattle farms and different questions asked by the consumers: cattle identification, sanitary aspects, food quality and trace back, animal welfare and environment. The Charter is organised at National and Regional levels. 50,000 farmers have signed this Charter, and a pool of 2,500 technicians have been trained to advise and accompany them. A recent evaluation of the Charter has been realised, which shows the necessity: to increase the communication around the Charter; to train frequently the technicians; to give farmers some tools to explain their action to the public; to organise specific advice to small farmers. These different aspects will be the main objectives of the next years.

Keywords: cattle breeders, advisory methods, good practices

Introduction

Cattle farmers are continuously making technical and economical decisions about their farms. To this purpose they need to harness their knowledge and their representation of the problem, and to implement discussions in different groups. Advisory services can help them. In France, different organisations (mainly with a professional or commercial vocation) are advising farmers. In 1962, a specific law created professional extension services. Since then, advisory practices and methods have been transformed, as the farmers' job also deeply changed. At the beginning, advice was mainly oriented towards technical subjects, and organised at a local and individual level. It progressively evolved towards economical topics, products quality, working organisation, quality of life... At the moment it is also trying to meet the societal demands : environment, animal welfare, food and animal traceability... In the same time, the farmers' demands and advisory methods became more and more diversified (CEREF-ISARA, 1993).

The target of this article is "the livestock farmer", mainly the dairy farmer. First, the context of technical change in agriculture will be described; then the different advisory methods will be studied, as well as the way they meet the expectations of cattle farmers. More details about the collective advisory and communication methods will be also developed. Finally the case of the French Charter of Good Practices for livestock farmers will be discussed (this Charter being at the moment proposed to all the French cattle farmers). This Charter is based on 35 practices, which are supposed to help farmers to do their job correctly and meet the demands of the general public. The objectives of this Charter, its organisation, the attitudes of the farmers about it, and some prospects for its evolution will end this presentation.

Technical change and advisory methods

Some explanatory factors for technical change in agriculture

Advice to farmers aims at helping them to make their decisions, and often to change their practices or their way of thinking. Sociology, psychology and pedagogy have been used to study different aspects of technical change, in order to explain it, and to help advisers to build their intervention methods. Below is our proposition to organise within four main categories all the factors that can explain changes (N'Sonde, 1998).

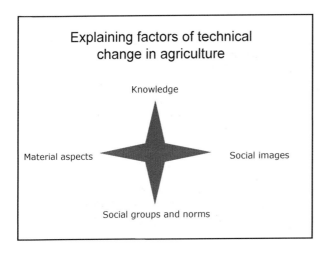

Figure 1. Explaining factors of technical change in agriculture.

- **Knowledge**: some practices might be explained by the lack of understanding of some biological aspects. For instance, if you do not understand what a leukocyte is, you may not make the link between mastitis and cell counts. If you do not know the presence of nitrogen in manure, you can not understand the link between manure and nitrates.... (Dockes *et al.*, 1999). Nevertheless, knowledge is not always necessary, nor sufficient, for the implementation of new practices. The perceptions farmers have of one phenomenon must be taken into account (Darre, 1985).
- The **representations** of the subject (Jodelet, 1993), for example the image of a practice (or the idea the farmer has of his job or of his social position) can explain his attitude about a subject. Soriano and Sens (1998) explain that representations constitute a way of organising one's knowledge with a coherence which is appropriate to each person. Kling-Eveillard (1999) uses the term of "attitude" emanating from the marketing sciences (Lendrevie & Lindon, 1990). Attitudes are ways of thinking, socially acquired capacities.

They refer to precise objects (working practices, status of the animal, relationship between man and animal, position of farmers in the society...). For instance, a farmer might think that identifying each animal can only be achieved in big and modern farms, and that his farm is not big and modern enough...

- images and part of the knowledge are elaborated into **social groups** (Darre, 1994), through discussions among farmers or within the family. Advisers bring information which are integrated into these debates (Lemery, 1994). The quality of the discussion with other farmers and with advisers explains a great part of the technical change. A farmer trusting an adviser more easily will follow his/her advice.
- finally, **practical aspects** are obviously very important (Jorion, 1980): it will not be easy to convince farmers to change for a practice difficult to implement. For example farmers usually find it very difficult to note their sanitary or fertilisation practices every day. A farmer whose back hurts will not easily agree with handling heavy buckets to clean his cows' teats...

The diversity of farmers' expectations

This chapter aims to demonstrate that French farmers constitute a very diversified population, according to their general attitude towards technical change as to their relationship with advice and advisers. Thus, the typologies of production systems are constituting precious elements for the orientation of the agricultural development (Perrot & Landais, 1993a; 1993b). In the same way, typologies of attitudes, which group together farmers expressing the same ways of thinking about a technical aspect (or about technical change and advice) are useful to conceive advisory actions well adapted to each public.

Two typologies of attitudes will be presented:
- The first one is sorting the farmers according to their opinion about technical change and advice (Birlouez *et al.,* 1994) ;
- The second one is rather dealing with social demands (such as the environmental and landscaped aspects). Our proposition consists in a typology based on the farmers' priorities and motivations with regard to the societal demands (Guillaumin *et al.,* 2004).

A typology about technical change and advice

The typology of farmers' attitudes we propose is based on discriminating 5 groups (Dockès *et al.,* 1999). The five types are named by two words, one describing their attitude towards innovation, and the other one their relation to advice.
- **The "independent innovators"**. They are usually clients of professional advisory organisations, but they often criticise them, finding them too conservative. They very much enjoy innovation and testing new practices or new productions syystems. They do not hesitate to take real economic risks to innovate. They remain well informed thanks to various and modern sources of information.
- **The "accompanied volunteers"**. They are always clients of professional advisory organisations, in which they often play an active role. They make their decisions on their own, after an economical reasoning. But they always take advice previously. They want global advice, taking the different sides of their farms into account. They are more and more interested in work organisation and in the way to improve their quality of life.
- **The "dependant moderates"**. They are clients of advisory organisations. They need and follow the advice to make all their decisions. They can not manage the economical aspects alone and are not always able to understand when advice is not convenient to their

production system. They could need a global and economical approach, but they usually do not ask for it.
- **The "autonomous moderates"**. They are usually not clients of advisory organisations, are not very interested in technical change and information. They are more interested in economic aspects, and in decreasing production costs. They are ready to discuss with other farmers and with technicians, but only occasionally, and if they are sure that their autonomy of decision will be protected.
- **The "isolated routiners"**. They usually see no adviser and only a few other farmers; they are quite alone to make their decisions, and do not like to change their working habits. The economic balance of their farms is often shaky, and they think that any change might be dangerous.
- They could benefit from individual aids taking their own preoccupations into account.

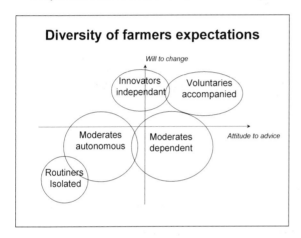

Figure 2. The diversity of farmers' expectations.

A typology of attitudes about the multifunctionality of agriculture (Guillaumin et al., 2004)

This typology was built to describe the farmers representations of multifunctionality, and their willingness to fit to social demand. Four types of attitudes were defined:
- **Interest in the producing function of agriculture**. These farmers aim at producing more and at increasing the competitiveness of their system, often in a prospect of marketing on the world markets. They consider the social demands as pure constraints.
- **Producing is the first function, but some environmental demands are integrated** into the practices of this second group of farmers (often in order to optimise technical and economical management of their farm).
- **The social and environmental functions are by products of the productive function.** These farmers try to maintain the family system, by taking what they consider to be good professional practices into account.
- **Services and agro-environmental activities are in synergy** with the function of "agricultural production", as real products. The priority of this type of farmers is to live in the country and to be accepted within a territory.

Advisory services: objectives and methods

Extension services have built a wide range of tools and methods to meet the diversity of farmers' demands.

Four types of aims for advice:
- **Strategic orientation**. It helps farmers to reason the way they set up their farms or an important change. The farm is considered as part of a system also involving partners, other farms, the countryside and the environment. This form of advice is dedicated to all farmers, but only a few times during their career.
- **Legal or management advice** which are mainly of interest to big farms.
- **Economic or technical-economical diagnoses and advice** aim at improving the functioning and management of the farm, taking the farming system into account. Such tools are quite frequent in France (Delaveau *et al.,* 1999), and are supposed to help all the farmers. In fact they mainly interest the "voluntaries accompanied", and the "autonomous moderate" ones, who pay attention to the link between technical and economic aspects. The development of these methods is often slowed down by the lack of spontaneous demand from the farmers, the lack of training for the technicians, and by the organisation of the advisory structures, mainly oriented towards technical advice.
- **Technical advice** is still the more frequent kind of advice proposed to farmers. It concerns their production ways (in order to improve their efficiency) but also their potential answer to society's demands (quality, pollution, animal welfare…).

Four kinds of methods for advice:
- **Individual advice** allows a direct dialogue between advisers and farmers; it can take every specificity of every farm into account, but is very time consuming for the technicians. This kind of advice is more useful when the adviser is able to understand the objectives and the points of view of each farmer, to adapt his advice and to make it be adopted.
- **Small group advice**, with one adviser and a group of about 10 farmers who regularly meet. It is particularly convenient for the "accompanied volunteers" who appreciate to debate among farmers, and with a technician.
- **Mass diffusion**, which mainly consists in diffusing information in the professional papers, is easy to realise, cheap, but rarely convincing enough to lead farmers to change their practices.

The next chapter will present a specific advisory method: **collective communication,** which aims at improving the productivity and the quality of advice by organising dialogue and interaction during meetings at a collective level.

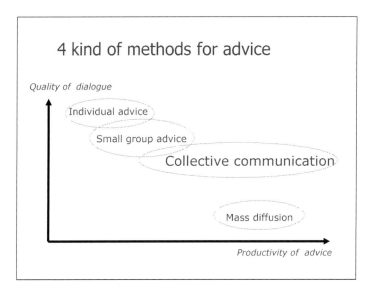

Figure 3. Four kind of methods for advice.

Collective advisory and communicative actions

During the last 20 years, the aims of the extension services and the farmers' expectations became wider and wider. At the same time, the financial resources of the services decreased. Hence new advisory methods had to be built (Dockès & Madeline, 1992; Dockès *et al.,* 1999). Collective advice and communication methods are managed as "projects", meaning that:

- They are based on four different steps (a maturation phase to decide the project, a preliminary study, some advisory actions, and finally the following up and evaluation phases);
- They have their own budget;
- They rely on different people and organisations, and thus need a specific partnership to be set up.

The preliminary study constitutes the occasion to correctly prepare the action by defining six specific properties of the project:

- its aims (allowing all the French farmers to follow good professional practices for instance);
- its targets (the young farmers, or the cattle farmers who do not already take part in milk performance recording for instance);
- its technical aspects (which practices are suggested to farmers, and why...);
- the best arguments to convince farmers (e. g. : easiness of the technical change, need to change to follow the expectations of the general public…);
- the partnership: every person in contact with the farmers must say the same thing in exactly the same way in order to be efficient;
- the kind of advisory methods and communication media to use. Usually, a combination of different kinds (documents, meetings, individual advice…) is necessary.

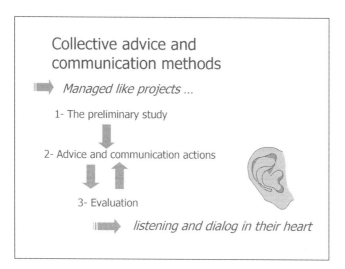

Figure 4. Collective advice and communication methods.

Advisory and communication methods

A schedule is defined at the end of the preliminary study. A coordinator is chosen and works for at least half of his/her time on the project. He/she coordinates the activities, manages the schedule, informs every actor of the project, communicates,... Different kinds of actions are generally used to interest and convince the farmers:

• Visual aids (logos, posters) are identifying the project, and reminding people of it;
• Written aids (press articles, specific letters or leaflets...), are giving information, and making people aware of specific problems;
• Small meetings are giving farmers the occasion to express their questions or difficulties; it is also the occasion for technicians to debate and convince the farmers to change their practices;
• Thanks to individual advice, the technicians can take all the specificities of each farmer and each farm into account. It is particularly necessary in case of major technical changes, or when the technician has to validate the farmers' practices (within the frame of a quality approach for instance).

Evaluation

Evaluation is part of the action as it constitutes the occasion to adapt it. The goal is not to judge but to develop ideas to improve the project. Different kinds of evaluations can be made, according to the step of the project in which they take place and to the objectives:

• Evaluation of project functioning: what do the different actors involved think of it, what do they suggest;
• Evaluation of the communication process: it aims to lead the actors to understand how the action is perceived by the farmers and technicians involved, which image of the project they have, and how to improve it;
• Evaluation of the means compared to the initial forecast;

- Evaluation of the realisations: did the actors of the project do what they were supposed to do, why?
- Evaluation of the technical changes, to follow up the adoption of the new practices by the farmers;
- Evaluation of the final impact compared to the objectives of the project...

Listening and dialogue are at the heart of project

One of the key factors of success for these actions is the ability to listen to farmers and to organise dialogues. Only people getting correct answers after having expressed their preoccupations and difficulties will be able to change. In individual advice, listening and dialogue are easy to organise, good training of the technician is sufficient. In the opposite, dialogue has to be organised in case of collective approach.

Preliminary and evaluative studies, as well as motivation studies are the best occasion to listen to farmers. Specific surveys will also help to understand the farmer's perceptions and expectations.

Participative meetings are a useful way to encourage dialogue among farmers, and between farmers and technicians.

Partnership and frequent discussions among partners also constitute a major factor of efficiency.

The Charter of Good Practices for livestock farmers

The Charter of Good Practices forlivestock farmers is a collective, professional and voluntary approach. It concerns all cattle farmers and is being set up since 2000. A collective advice and communication project has been organised to encourage farmers to adhere to the Charter. An evaluation of its implementation was realised in 2002.

The context and the objectives of the Charter

The different crises which have occurred during the recent years in cattle production have changed the behaviour of the consumers and public opinion. Thus the professionals in charge of the breeding sector decided to set up the Charter of Good Practices in Cattle Breeding in order to restore a good relation between the public opinion and the farmers, and to promote the occupation of "breeder".

The Charter was created by the National Bovine Federation and the National Milk Producers Federation, and aims at restoring the links between the general public and the farmers. The idea was to define the "good practices" of a breeder and to teach them to the consumers. The Charter also aims at progressively improving the cattle breeders practices and production modes.

The Charter has thus a double objective:
- To collectively ensure the confidence of the consumers by respecting and publicizing good practices;
- To individually propose to the farmers a way to progress and improve their practices.

The contents and the organisation of the Charter

The Charter proposes that each livestock farmer voluntarily undertake about thirty interrelated practices within seven fields:
- cattle identification;
- sanitary qualification;
- use of medicines;
- healthy and controlled food;
- hygiene and dairy production;
- animal welfare;
- environment and farm access.

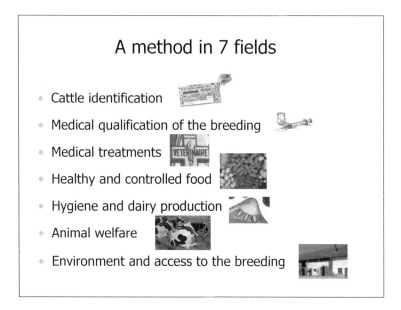

A method in 7 fields

- Cattle identification
- Medical qualification of the breeding
- Medical treatments
- Healthy and controlled food
- Hygiene and dairy production
- Animal welfare
- Environment and access to the breeding

Figure 5. A method in 7 fields.

The Charter is set up in the French regions. Its implementation and monitoring are under the responsibility of regional Steering Committees which rely on regional "project leaders". Some technicians of professional organisations of Accompaniment and Validation are specifically trained for its implementation on the farms. Commercial or extension services are also involved (dairies, milk performance recording organisations, chambers of agriculture..).

At a national level, the Charter is organised around a National Steering Committee and the French Livestock Institute. The Steering Committee is composed of national livestock farmers' organisations and makes decisions related to the management of the charter (structure and organisation), its operation and its evolution. The French Livestock Institute leads the technical Committee in charge of the working out of these proposals.

Farmers are joining the Charter voluntary but have to sign a specific document. The technician in charge of the farm visits must have been specifically trained and must be "approved" by the regional Committee.

The functioning of the Charter relies on the following steps:
- Self evaluation by the farmer;
- Validation of his adhesion to the Charter an agent of extension services or economic organisation;
- Validation visits every two years;
- Monitoring of the organisation based: on the one hand, on an internal monitoring; and on the other hand, on the intervention of an independent expert approved to carry out an "external control".

The way the livestock farmers are supported

The farmers are informed of the existence and of the principles of the Charter through collective media like press articles, specific (or non specific) meetings, and individual contacts with technicians and other farmers.

When a farmer has decided to join the Charter, he must ask for the intervention of a technician (or the technician may suggest the farmer to become involved in the Charter).

During the validation visit, the intervention of the technician aims to evaluate the situation together with the farmer, and to compare it to the different rules of the Charter so as to identify eventual progress margins. The goal is not to sanction the breeder, but to help him to progress.

This visit can thus be compared to advice, comparing the farmer's situation to the Charter's targets. It constitutes a privileged moment for dialogue. The technician must observe the farmer's practices but above all he must devote himself to listening and to discussing with the cattle breeder.

The Charter is a way to progress based on a dialogue between technician and farmer, and on the commitments of the breeder to improve his practices. The following validation visits should be carried out by the technician who validated the initial adhesion.

It is essential that the breeder is followed up between the two visits (by the technician of OPAV during his daily work or by other advisers); this way he can progress on the different headings of the Charter. It is particularly necessary for breeders undertaking to change some practices which they have not already adopted at the very moment of the adhesion. Of course the success relies on the technicians, but it also depends on the actions set up to develop the Charter (financial subsidies, meetings, information, …) at the regional level.

The appreciation of the Charter by the farmers

During the evaluation of the Charter's implementation in 2002 (Kling & Hédouin, 2003), 209 dairy farmers were individually interviewed on their farm. Out of these 209, 101 had already adhered or were committed to the Charter.

Most of them knew that the Charter existed, but they did not know its content very well: it was often confused with other quality approaches, and the one implemented by their dairy enterprise in particular. The milk performance recording organisations often played an important role to inform the farmers.

The farmers's general knowledge was often limited to the name and to some elements of technical contents of the grid. Few farmers knew the origin and the objectives of the Charter, thus explaining their disappointment when learning there was no direct financial return to expect.

The farmers' opinions on the interest of the Charter were generally divided: some of them considered it as a way to communicate with "the outside world", and to value the occupation of farmers; others used it as a personal tool to progress and improve their own practices.

It should be noted that these interests were exactly the two national objectives of the Charter. Its image seemed to be better in the regions which were already engaged in local quality approaches for a long time.

Some farmers also told about the valorisation of the efforts made to adhere that they expected: financial valorisations for some, or communicative ones (towards the consumers) for others.

The reactions to the technical contents of the Charter varied a lot. They often seemed to be perceived as relevant by the farmers, who recognised them as good professional practices, more than right answers expressed for consumers.

The actions most frequently quoted as being difficult to set up were the following:
- Systematically writing down the medicines in the medical notebook, and keeping all the prescriptions;
- isolating the introduced animals;
- cleaning and improving the accessibility of the buildings.

Actually the most difficult practice to set up is the first mentioned above.

In fact, the four types of explanatory factors for technical change were expressed (knowledge, image, discussions in social groups and with technicians, and practical aspects of the practices). The hindrance and the motivations expressed by the farmers made it possible to describe three different **attitudes** about the Charter:

First type: really convinced of its usefulness

Few farmers belonged to this type. They were rather sensitive to the opportunity to communicate on their occupation to improve its image. They found the Charter as complementary to the other quality approaches in which they were already engaged, and in which they believed; they also perceived it as a progress tool. They belonged to the groups of "innovators independent and volunteers accompanied" that we described in the previous chapter.

Second type: neither really for, nor really against

They were the most numerous. Generally they adhered to the Charter because it did not require many changes on the farm, or because it only brought few new constraints, or because the technician convinced them to adhere, or because the adhesion gave access to financial subsidies.

It is important to underline that many breeders thought that the Charter would sooner or later become obligatory. That is why they preferred to adhere now. They often belonged to the groups of "moderates" of the previous typology.

Third type: reticent

There were very few, and were reticent about their commitment to the Charter just as they often were reticent to commit themselves in other quality approaches. Several reasons were evoked by them: no interest, too many constraints, refusal of principle. They belonged to all categories of farmers.

This typology shows that there are few "really convinced of the interest of the Charter" but also few reticent people. The majority hesitates to integrate an approach about which they do not know much in terms of specificity and interest.

Farmers who do not adhere to the Charter can be found within the three types. Some even said they were interested and would have like to adhere, but were not informed of the existence of the Charter, or knew it and were astonished not to have been contacted by the technician yet. This shows the existence of a potential of farmers to contact, then to support towards the adhesion to the Charter.

We will now develop the way adhesions to the Charter may be improved.

The way to improve adhesions

Some situations and characteristics appeared to be determining to explain the success of the Charter in some areas and, on the contrary, the difficulties which slowed down its implementation.

The first positive aspect for the Charter's development was the strong implication of the "dairy" actors. The dairy industries realised nearly 75 % of the validation visits. Doubts persisted however on the importance they put on the Charter's rules at the time of the visits, and on the way they presented it to the farmers. In some areas, extension services were also deeply involved in the Charter.

In addition, these regions shared some characteristics with the areas which have the highest rate of adhesions at the moment:
- A "clear" and formalised organisation;
- A strong conviction of the regional "project manager" which set up the Charter with or without the implication of local ("departmental") organisations;
- An coordinating effort of the regional "project manager" to gather the local organisations;
- The presence of farmers in charge of the development of the Charter in the regions, and of the sustaining of the regional "project manager".

In the opposite, evaluations sometimes raised difficulties emanating from the organisation around the Charter: lack of organisation, dialogue and information between actors, lack of coordination between organisations, heterogeneity in the interventions of the technicians and in the local operation (in the organisation, the implication of the partners, the motivation and the results). These problems slowed down the development of the Charter in some regions or departments.

It is thus necessary to reduce the heterogeneity between the actors presenting the Charter, in order to take advantage of their complementarities and to facilitate new adhesions made in good conditions.

The following improvements should therefore be implemented:
- a clear definition of the roles of the technicians,
- a true coordination between the "departments" and the local actors,
- better training and exchanges to facilitate the action of the technicians,
- more information about the difficulties met by farmers in order to ensure that all commitments will be respected.

Beyond the specific organisational problems of each area, some important questions are still remaining such as the implication of all the actors involved in meat production. This question will have to be solved to improve adhesion.

Among the listed difficulties, the motivation of the farmers and of the actors of the Charter (especially the technicians and professionals in charge of the organisations) seems to be one of the significant points for the dynamics of the regional actions.

The implication of the professionals in charge of the organisations is sometimes perceived by the actors of the Charter as not being strong enough. For several reasons, they should be more deeply involved in the Charter:

- in terms of principles, the Charter is a professional approach, which aims at promoting the occupation of breeders, complementary of the products' quality approaches;
- these professionals should be ambassadors of the Charter;
- people working in the different organisations should be motivated to make the Charter their priority.

A communication campaign is to be conducted:

- towards the actors of the Charter (local technicians, organisers…). It seems important to motivate all the partners. This was awaited by many people met during the evaluations.
- towards the farmers: they must know the Charter as well as possible and distinguish it from the other quality approaches. Those who know little about the specificities of the Charter will not be able to speak about it. It will also be difficult for them to convince some visitors of the interest of the Charter. In the areas with the highest level of knowledge, there will probably be a positive effect on the adhesion of new breeders. Beyond knowledge, it is essential to motivate breeders to make them understand the "collective" interest of the Charter. It is thus necessary to speak directly to the breeders about it, and to do it regularly, by informing them on the way it is evolving.

The communication must also be a way for farmers to promote their profession and their good practices. The system of the "breeding panel" which is already in place in many areas is a first conclusive possibility since it motivated other farmers to adhere to the Charter. To this purpose, a specific document was written in order to help them to present their occupation and to explain why they adhere to the Charter.

Conclusion

The Charter of Good Practices is a good illustration of the different approaches for advisory services dedicated to livestock farmers.

Its various aims are the following:

- to help the farmers to explain their job to the general public, and to meet its main expectations (environment, animal welfare, animals and food traceability);
- to help them to improve their practices;
- and to help them to improve their income (by improving product quality) and their working conditions (in terms of personal organisation).

Different advisory methods are thus implemented:

- individual advice, which plays the main role in the Charters' organisation;
- collective advice to inform farmers of the existence and interest of the Charter;
- financial incentives: some subsidies are now linked to the implementation of the Charter.

The Charter is organised as a project, with national and regional levels. Specific studies, and especially the recent evaluation of the implementation of the Charter, aim at organising and improving its functioning. The main issue is now to communicate with efficiency about and around the Charter, to convince all the French livestock farmers to join it and to enable them to do it thanks to advice and subsidies. At the moment more than 100,000 farmers are involved in the Charter, the target being to reach some 200,000 farmers within three to five years.

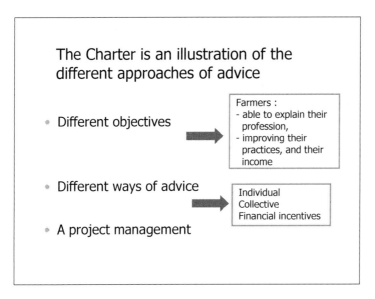

Figure 6. The Charter is an illustration of the different approaches of advice.

References

Birlouez E. & A.C. Dockes, 1994. Construction de services d'appui technique destinés aux éleveurs laitiers des Pays de la Loire. Institut de l'Elevage, GIE Lait Viande des Pays de la Loire. Document de travail, France, 50pp.

CEREF - ISARA, 1993. Les attentes des agriculteurs en matière de conseil. ANDA.France, 35pp.

Darre J.P., 1994. In Pairs et experts en agriculture. Eres. France, 7-14

Delaveau A., C. Perrot, E. Tchakerian & J. Véron, 1999. La cohérence des techniques fait le résultat économique. Renc Rech Ruminants 1999, 6, France, 3-10.

Dockès A.C., M. Lenormand, F. Kling-Eveillard & Y. Madeline, 1999. Vers l'intégration des différentes démarches de conseil aux éleveurs. Renc Rech Ruminants 1999,6, France, 55-61.

Dockes A.C. & Y. Madeline, 1992. L'Ingénierie des actions de conseil collectif. L'exemple de l'opération Fourrages-Mieux. INRA, Etudes et recherches sur les systèmes agraires et le développement, France, n°25, 1-42

Guillaumin A., D. Bousquet & A. Villaret, 2004. Multifonctionnalité de l'agriculture: identification des demandes locales et de leur acceptation par les agriculteurs. INRA, Cahiers de la Multifonctionnalité n°7, France, 125-136.

Kling-Eveillard F. & C. Hédouin, 2002. Charte des Bonnes pratiques en Elevage. Synthèse des évaluations Nationales. Institut de l'Elevage, Paris, France, 55pp

Jodelet D., 1989. Les représentations sociales, PUF,France, 350pp

Lemery B.,1994. In Pairs et experts en agriculture, Eres. France, 91-116.

Mendras H. & M. Forse, 1983. Le changement social, tendances et paradigme. Armand Colin, France, 386pp

N'Sondé Senga L., 1998. La dynamique du changement de pratiques de traite en élevage laitier. Mémoire de DEA, Université de Nanterre, France, 81pp

Perrot C. & E. Landais, 1993a. Exploitations agricoles: pourquoi poursuivre la recherche sur les méthodes typologiques. Les Cahiers de la Recherche-Développement, 33, France, 13-23

Perrot C. & E. Landais, 1993b. Comment modéliser la diversité des exploitations agricoles? Les Cahiers de la Recherche-Développement, 33, France, 24-40

How environmental problems are addressed to farmers – pyramid model, research, knowledge transfer, practices and attitudes

Abele Kuipers[1], Karin de Grip[2] and Paul Galama[3]

[1] *Expertisecentre for Farm Management and Knowledge Transfer, Agro Business Park 36,
6708 PW Wageningen, The Netherlands*
[2] *Communication and Innovation Studies, Wageningen University Hollandseweg 1, 6706 KN
Wageningen, The Netherlands*
[3] *Animal Sciences Group, Wageningen University and Research Centre, Edelhertweg 15,
8219 PH Lelystad, The Netherlands*

Summary

In the 1990's the public extension service was semi-privatised. The Ministry of Agriculture continued to support extension activities in areas such as environment and nature management. The know-how development and transfer related to environmental activities in the dairy sector was organised like a pyramid model. This set-up was later copied in the grain and arable sectors. From top to bottom we find research farm(s), discovery farms, field-demo-farms and the large group of "average" farms. As a research farm, The Marke was founded in 1990, where management practices were examined with the aim of reducing mineral losses. Later, 17 discovery and 175 field-demo-farms were established to further test and gain on-farm experience of the various management practices. Traditional reports, field days, farmer meetings and study-groups were organised to support the dissemination of the experience gained. Keywords in the approach appeared to be: clear goals, farmer-to-farmer exchange, objectivity and an open attitude. However, studies indicated that these activities were not enough to really encourage the "average" farmers to adapt to more environmentally friendly practices. A Mineral-management Liaison Service was set up in 2002 to stimulate the knowledge exchange further and to experiment with demand driven extension. 80,000 farmers could request a voucher worth € 250,- for free to buy a knowledge product. These products were exposed in the "knowledge shop" on website. Farmers could also join a study group in order to articulate their needs together and to buy a "knowledge product" collectively. Trained farmers co-ordinated these groups. The Liaison Service also developed a quality system, using client satisfaction studies to contribute to a transparent knowledge market. Reflections on these initiatives are presented in this study.

Keywords: knowledge transfer, pyramid model, practices, attitudes, demand driven initiative

Introduction

In the late 1980's, the extension service in The Netherlands was privatised. Part of the service was taken over by feed companies and private consultancy agencies. However, the Ministry of Agriculture continued to support and initiate extension initiatives around themes of public interest, such as environment and food safety. We want to illustrate the structure in which the public budget is devoted to research and extension initiatives related to environmental issues.

In The Netherlands, the reduction in mineral losses has been an area of utmost importance to meet environmental policies (Kuipers *et al.*, 1999). The structure of applied research and

knowledge transfer in the environmental area can be best visualised by a pyramid model (see Figure 1). The philosophy behind this model is:
1. Knowledge development takes place at experimental research farms
2. The research findings are further developed and demonstrated in the field at a limited group of discovery farms; emphasis is on development
3. A larger group of demonstration farms are used to implement and demonstrate the farm practices needed to achieve the required results; emphasis is on demonstration
4. A farmer-to-farmer approach is used to try to reach most of the farmers. The idea is: let farmers tell farmers the news about topics, like environment.

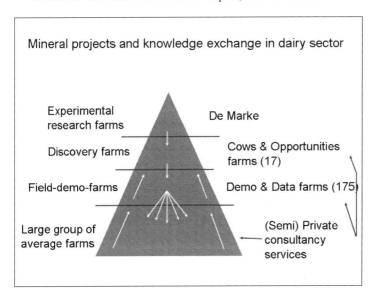

Figure 1. Mineral projects and knowledge exchange in the dairy sector.

In this contribution we will share some research results and experiences that are gained by applied research and reflect on the extension approach of using the pyramid model. Also, experiences with the so-called Liaison Service are discussed. This Service was set up to deal with knowledge transfer to the large group of commercial farmers and to experiment with demand driven extension work.

Experimental and demo-farms

De Marke experimental research farm was built in 1990 to enhance practical knowledge about mineral (nutrient) management (see Figure 2). This experimental farm is supported with scientific knowledge in the animal husbandry and environmental fields by the Research Institutes in the country.

Figure 2. Environmental research farm De Marke.

Ten years of applied research resulted in successes and failures in the management practices tested. Some of the successes and failures are listed below:

Management *successes* at De Marke	Management *failures* at De Marke
• 40 % less N fertilization, the same crop yield	• Siësta grazing too laborious
• Grass under maize appeared to be an environmental friendly practice	• Closed sloping floor too slippery (used for reducing ammonia emission to air in barn)
• 15 % crude protein in ration is sufficient	

The **discovery farms,** named **the Cows & Opportunities farms,** were selected to further develop and gather on-farm experiences with mineral management practices developed at De Marke and other research farms. The research goal at the discovery farms was to reach the EU-environmental targets within 3-years. This was an obligatory goal to the farmers. The farms are listed in Figure 3. The 17 dairy farms were spread over the country and over soil types.

The **field-demo-farms,** named the **Demo & Data Farms** also tried to bridge the gap between the experimental and commercial average dairy farms. These farmers were requested **to try** to reach the environmental targets. It was not a must to reach these targets. Support was less intensive than at the discovery farms. 175 dairy farms participated in this project.

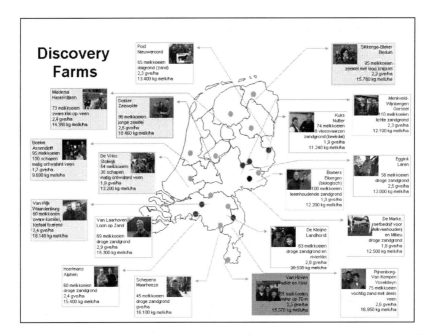

Figure 3. Discovery farms.

Experiences

The environmental practices which were most popular on the discovery and field-demo-farms are listed below. Especially the usage of manure got a lot of attention.

Environmental practices most used:
- Better utilization of manure
- Less fertilizer use
- More clover
- Less protein in feeds
- Less grazing
- Less young stock
- Investment in quota and land
  (discovery farms: 6 years + 124 tons of quota + 8 ha's)

In order to better understand the role of environmentally sound practices within the total of farm management practices, a survey was performed with the field-demo-farmers (Kuipers *et al.*, 2001). They were asked to rank 10 management practices in order of importance. As can be seen below, mineral management was not considered as a priority on farm level. However, many farmers added that mineral management is in fact an integral part of the management of the farm.

Management practices ranked in importance:
(1 = most important; 10 = least important)

- **Feeding**   **3.1**
- **Economics**   **3.5**
- Grassland management   3.8
- Animal health   3.9
- Milking   5.1
- **Mineral management**   **6.0**
- Calf rearing   6.3
- Labour organisation   6.8
- Animal breeding, selection   7.1
- Machine work   9.2

Results show that the effects of mineral management practices on the discovery and field-demo-farms differ from the average farms. The N-surpluses of the front-runner farms (discovery and field-demo-farms) were, as expected, considerably below the commercial average dairy farms (see Figure 4). The surpluses on the front runner farms approached the EU-targets. However, large differences existed between farms within groups.

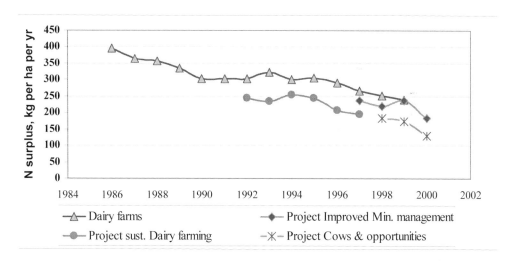

Figure 4. Mean N surpluses of different groups of dairy farms.

Various initiatives were undertaken to organise workshops to discuss plans and results with parties, like government officials, feed companies and water quality agencies. De Marke received 3,800 visitors per year. The 175 field demo farms were visited by about 8,700 people over the last 3 years. In the winter season of 2002-2003, lectures were given at meetings which were visited by a total of 1,500 farmers. The concept of the meetings was that a discovery farmer shared his experiences with colleagues visiting the meeting (thus a discovery farmer tells the story).

The farmers were asked about their experiences in participating in these research and demo projects. The most common remarks were:

- They are proud to be a front runner; they profit from interaction with research institutes and the extension service;
- They love to have colleague farmers visiting their farms;
- They also received criticism from colleague farmers about helping the government to reach the environmental goals.

The researchers and advisers who are supporting the farms stress the importance of the following approach: we let innovative farmers tell and show their neighbours, friends and colleagues what they are doing, why they are doing it and what the technical, financial and environmental results are. Scientists and advisers are only the facilitators in this process, they are not the major actors. Excursions and study groups are approved activities where farmers learn from farmers. A survey among progressive farmers, the Cows and Opportunity farmers showed that their "change of thought" started in an early stage while visiting active colleague's farms and when making the first calculations about application of minerals (Beldman & Doornewaard, 2003). Experiences outside agriculture can also be very influential on the way of thinking. Publicity in agricultural magazines etc. can give additional support.

Keywords for an effective communication-strategy appeared to be: objective, reliable, open and positive. Objective: no hidden agenda, no propaganda. Reliable: sound science. Open: tell and show everything, successes and failures. Positive: every problem is an opportunity and a challenge.

Lessons learned

In conclusion, some **lessons learned** during the execution of the projects are presented below. But it should be emphasised that this structure of knowledge exchange by a pyramid model is only used for the environmental theme in the Netherlands. It may be considered as an experiment. The pyramid model has its restrictions.

Lessons learned:

1. Discovery and Field–Demo projects create a lot of interest of public parties
2. Awareness raising is a very complicated process
3. Strong overall co-ordination is needed between different initiatives to create coherence
4. The approach in projects is often research oriented; the communication strategy is less developed
5. The network involved in the projects is limited; intermediaries should play a greater role
6. Knowledge about mineral management should be part of the total farm management approach
7. But still: the approach was not really effective in fostering a significant change in management practices at the large group of commercial average dairy farms

---

How to deal with knowledge exchange with the large group of farmers: an initiative

The so called Liaison Service was set-up to deal with knowledge transfer towards a large group of farmers. The Service is a three year initiative, and became operational in 2002. The Ministry of Agriculture finances this initiative with a budget of € 25 million. They looked for an independent organisation to execute this initiative. The main aim of the initiative is to exchange knowledge on mineral management with a large group of farmers.

At the same time, their was the request of the Ministry to organise knowledge exchange differently and experiment with demand-driven knowledge exchange. The idea behind the Liaison Service was to channel money to be used for knowledge exchange activities directly to the end-users. In this case farmers could receive a voucher to spend on knowledge products in a kind of knowledge shop.

The Liaison Service organised four main activities to bring the knowledge demands of farmers and the knowledge supply of service organisations together. These activities are also considered as the four elements of the experiment:

1. A **voucher**. Farmers could request a voucher of € 250. This kind of coupon or artificial money gave them purchasing power to buy a "knowledge product". Farmers could buy different kinds of knowledge products, such as individual advice, supporting software, study group activities, courses on mineral management, excursions and books.

Figure 5. Voucher.

2. A **quality system.** This system was set up to derive a selection of knowledge products from the "knowledge shop". A board of experts used certain criteria for this selection. Client satisfaction studies were also part of the quality system. In this way the quality of knowledge products ought to be guaranteed.
3. **New study groups of farmers**. These study groups were co-ordinated by farmers, who were trained by the Liaison Service to become a study group co-ordinator. The idea was that these co-ordinators should be able to guide the group to get the knowledge demands clear. Then, the group of farmers could spend their vouchers together. This approach would probably be more efficient and effective than individual advice, etc. Also existing study groups could apply for a group voucher, which they could spend on a knowledge product.

4. A **knowledge shop.** This shop was developed to increase the transparency of knowledge products and allowed farmers to compare knowledge products. A website and leaflet were used as media for the creation of a knowledge shop.

This initiative was studied to learn more about this new approach to organise knowledge exchange. Three research questions were formulated, to guide the research:

1. To what extent do different parties recognise the aims of the Liaison Service?
2. To what extent do the elements [voucher, quality system, study groups, knowledge shop] of the Liaison Service support demand driven knowledge exchange?
3. Is there a future role for an Independent Service like the Liaison Service?

In terms of a methodology, in-depth semi-structured face-to-face interviews and structured telephone interviews were used among knowledge suppliers, study group co-ordinators, farmers, government officials and project team members. Participant observation was also conducted during various study group meetings, seminars and project team meetings. The main findings are listed below.

Findings

The main quantitative findings about the elements of the Liaison Service are the following (De Grip & Leeuwis, 2003). The target group of the Liaison Service is the 80,000 farmers and horticulturists in the Netherlands who are obliged to use the mineral management administration system (MINAS). These farmers and horticulturist could request a voucher from the Liaison Service.

The following results were achieved:
- 35,500 vouchers distributed to farmers
- 23,600 vouchers spent by farmers (Figure 6)
- 53 study group co-ordinators started 161 study groups
- 850 existing study groups requested and spent group vouchers

Accordingly, it became clear that most of the vouchers are spent on individual advice, namely 76 % (= about 18,000). Another 11 % (= 2,600) is spent on supporting software, 8 % (1,900) on the specially set up study group programs about mineral management. And a small part of the vouchers is spent on regular existing study group programs, excursions and books.

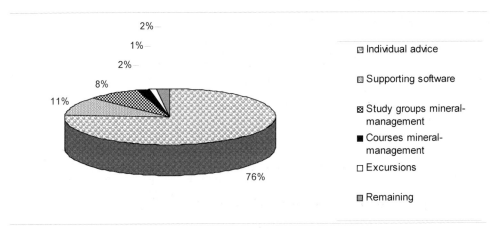

Figure 6. Kind of purchased knowledge products (n = 23,600).

Furthermore, it was found that only parties that were closely related to the Liaison Service (such as project team, knowledge supplier, study group co-ordinators) were familiar with the aims of the Service. In general, the farmers as target group did not really have an idea what the Liaison Service tried to establish. Of course, they received the information about the vouchers, but the whole idea that farmers should consciously think of their knowledge demand and critically think about a knowledge product was unfamiliar to most farmers.

It is also interesting that 85 % of the interviewed farmers (n = 39), did not perceive a mineral problem on their farm. More than 50 % of the respondents (n = 33) were self-confident about having enough information and knowledge to handle the mineral management issues at their farms properly.

Vouchers

About the first element of the Liaison Services, the vouchers, it was found that vouchers mainly activated the existing network around farmers, and only indirectly the farmers themselves. Many knowledge suppliers informed their regular clients about the vouchers, and the opportunity to spend the vouchers with them. About 80 % of the vouchers were spent in existing and familiar relationships between farmers and knowledge suppliers.

It also became clear that vouchers were mainly perceived as financial incentives. About 40 % of the interviewed farmers indicated that vouchers were useful because it gave a reduction on their existing post for extension and advisory costs.

The findings also showed that farmers hardly had a clear, well articulated knowledge demand about mineral-management. Farmers seem to perceive mineral management as an integrated part of the daily farm management activities.

A last finding about the vouchers is that it caused enormous administrative and bureaucratic procedures, especially for knowledge suppliers.

Quality system

Concerning the quality system, there were some clear indications that the quality system contributed to an increased quality of knowledge products. 40 % of the offered knowledge products were new products. Also, 34 % of the knowledge suppliers indicated that the quality of their knowledge products increased, because the products were checked by the board of experts.

Most farmers were satisfied about the product they bought with the vouchers. However, it also became clear that information about the quality of knowledge products is not so important to farmers in making a choice how to spend their vouchers. This choice is much more based on the familiarity with the knowledge supplier and the practical product-applicability on the farm, than on quality/price comparison.

Study group and study group co-ordinator

The most newly formed study groups were actually existing study groups or part of an existing study group. Also, most of the participating farmers already knew the study group co-ordinator. Accordingly, study group co-ordinators mainly arranged what kind of knowledge product the study group bought with the vouchers to be used at the meetings. Therefore, it was not a clear and structured process of demand articulation including all of the farmers of the study group. However, during the study group meetings there was room for specific questions and demands of the individual participating farmers (see Figure 7).

It also became clear that study group co-ordinators were not really well trained to guide the process of demand articulation. This aspect needs some more attention. Now, co-ordinators were mainly busy with administrative and logistic procedures.

Figure 7. Study group on tour.

Knowledge shop

Concerning the last element of the Liaison Service, the knowledge shop, it became clear that most farmers were not really interested in the transparency of knowledge products. About 85% of the interviewed farmers had a look at the leaflet, but did not really use it to make a choice on how to spend the voucher. Only 21 % of the farmers had ever visited the website, but there are no indications that they used this medium to choose their knowledge products. Farmers relied much more on existing relationships with their knowledge supplier(s).

Future role of Liaison Service

Related to research question three, most respondents said they would desire an Independent Service for the agricultural sector. An inventory about a possible future role of such an Independent Service indicated that this Service should contribute to create transparency about knowledge products. This seems a bit paradoxical given the fact that respondents did not really use the website and leaflet which were developed to create transparency about the knowledge supply side. It might be an indication that transparency should be organised differently.

Reflections

Reflections on the main findings of this study lead to three main points:
1. There seems to be very little autonomous urgency from farmers to adapt their farm practices related to mineral management. The main reason for this was that farmers are still waiting for clear policies about the required mineral targets. And farmers are tired of mineral management, they show an 'allergic' reaction towards the subject. In this context,

it has been very hard for the Liaison Service pro-actively to stimulate farmers to make changes related to mineral management on their farms.

2. Financial demand dominates substantive demand. It is mainly the financial incentive, i.e. the voucher, that makes farmers buy knowledge products. The reaction of farmers is not based on an autonomous clearly articulated knowledge demand.

3. It is doubtful whether demand driven knowledge exchange is really established through the Liaison Service initiative. Conversely, it is mainly the supply side of the knowledge exchange, i.e. the knowledge suppliers which became more activated.

Despite the fact that the Liaison Service did not meet all it was aiming for, and that it seems very difficult to establish a truly demand driven knowledge exchange, the conclusion shows that it is worthwhile to continue and improve this experiment. We conclude this because the research taught us that:

- vouchers contribute to the activation of the network around farmers; and indirectly, many farmers became activated to do something about mineral-management.
- the respondents desire a kind of Independent Service in the agricultural sector dealing with knowledge exchange, with a possible role of creating transparency in the knowledge supply side.

There are two other reasons which legitimate this conclusion, namely:

- that more time is needed to learn about an initiative like the Liaison Service and to develop and organise a clear and effective structure for knowledge exchange.
- that the government is (partly) responsible to search for mechanisms to organise knowledge exchange around publicly desired themes, such as mineral-management. So, they might as well continue to support this initiative in order to improve knowledge exchange with the large group of farmers.

Conclusions and Recommendations

Finally, some recommendations were formulated for improvement and optimisation if an initiative like the Liaison Service is to be continued. The most important one we like to mention here, is that we recommend that more attention should be paid to **the concept of demand driven knowledge exchange**. Realising that practice is more dynamic and complex, we identified four iterative steps which could guide the process to establish more demand driven knowledge exchange:

✤ **Step 1:** *Activation*
Farmers should be activated to try out new information channels, like participating in a study group or invite an advisor. The voucher can be a financial mean to do so.

✤ **Step 2:** *Awareness raising*
When a farmer tries out a new information channel, he can become aware of a problem on his farm.

✤ **Step 3** *Demand articulation*
When in the second step a problem is identified by the farmer, specific demands about this can be articulated.

✤ **Step 4:** *Demand driven knowledge exchange*
If the demands are clear for the farmers and put forward to the knowledge supplier, there is really a situation of demand driven knowledge exchange. This assumes that knowledge suppliers are able and willing to accommodate these demands.

If these four steps are kept in mind in the development and organisation of an initiative like the Liaison Service, knowledge exchange with the large group of farmers might be really based on what farmers demand. In that way, knowledge exchange can contribute more effective to the adoption of publicly desired farm practices.

References

Beldman, A. & G. Doorneveld, 2003. From seeing advantage to strategy (in Dutch: Van kwartje tot strategie). Cows & Opportunities report nr. 14.

Grip de, K. & C. Leeuwis, 2003. Lessons about demand articulation - experiences with the Liaison Service. Expertisecentre for Farm Management and Knowledge Transfer, report nr. 16.

Kuipers, A., F. Mandersloot & R. L. G. Zom, 1999. An approach to nutrient management on farms. Journal of Animal Science, Vol. 77, suppl. 2 / Journal of Dairy Science, Vol. 82, suppl. 2, 6pp.

Kuipers, A., B.W. Zaalmink & D. Kuiper, 2001. Usage and acceptance of nutrient management programmes by dairy farmers in Demo & Data Farms project. Expertisecentre for Farm Management and Knowledge Transfer, report nr. 3.

Extending the University to Oregon livestock producers

James R. Males

Oregon State University, Department of Animal Sciences, 112 Withycombe Hall, 97331-6702 Corvallis, Oregon, USA

Summary

Extension is one of the three components of the Land Grant Educational System in the USA. A successful program recently implemented in Oregon is the Beef Quality Assurance program. A dichotomy in potential clients is observed. Also the extension faculty is increasingly asked to take sides. Maintaining an active research involvement will keep the faculty progressive.

Keywords: extension, land grand educational system, challenges, choices

Introduction

Extension is one of three key components of the Land Grant Educational System in the United States. The other two are research and teaching. The relationship between these three missions of the Land Grant University system vary among the various states but an ideal relationship is demonstrated in Figure 1. Extension in the United States is funded with federal, state and county (local) funds. The portion of each and how they are used to pay salaries and cover support costs varies with each state. Extension personnel are employees of each state Land Grant University. There are people in Extension with state wide and local responsibility. The actual design of this also varies among the states.

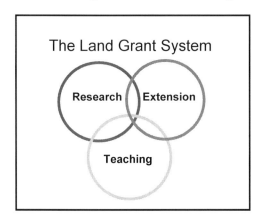

Figure 1. Ideal relationship among research, teaching and extension.

At Oregon State University we have an obligation to not only extend information to Oregon producers but also to those beyond its borders. Our mission in the Animal Sciences Department at Oregon State University is to serve animal agriculture and the people of

Oregon, the region, nation and world through research and education. Specifically addressing Extension education, our goal is to provide education that serves the needs of Oregon's animal industries and helps them respond to societal concerns. In Animal Sciences at Oregon State University we have tried to structure our faculty to fit in the relationship depicted in Figure1.

An example of a very successful Extension program we recently implemented in Oregon is Beef Quality Assurance. This educational program is designed to provide certification to producers who successfully undergo the training. The purpose of the program is to enhance beef meat quality by addressing management issues from conception to the plate. The program consists of a four to five hour training program put on by the Extension Beef Specialist and the Extension Veterinarian. Producers are encouraged to consider their facilities, breeding programs, nutritional programs, and their overall health plan and treatment methods. Accurate record keeping is stressed as an important part of any quality assurance program. The bottom line of the program is that all producers have to share in the responsibility to ensure that they are doing all that can be done to meet the high standards the public expects in food safety. After completing the program, producers can then take an on-line test. If they successfully pass it, they are issued a certificate that they are BQA trained. To date we have had 630 Oregon producers successfully complete the testing. Just as significantly, we have two "branded" niche market beef programs in the state requiring this certification of their cooperating producers. The Oregon program has been identified by our National Cattleman's Beef Association as one of a handful of model state programs that other states should emulate. We have also recently started similar programs in both sheep and dairy.

What does the future hold?

As we look at the future of the animal Extension program in Oregon we see a number of challenges, but the two largest are identifying our audience and managing contentious issues. Extension as an organization was traditionally agriculture and natural resource based with a strong family and youth component through the 4-H program, and was designed to serve primarily rural communities. As our agricultural production has shifted to more industrialized production we see more of a dichotomy in our potential clients (Figure 2).

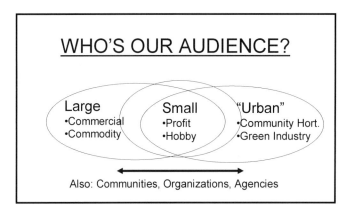

Figure 2. Range in audiences that are demanding extension education.

On the one side we have large commercial farms wanting very technically competent, cutting edge information. They are utilizing consultants more and more to handle their needs. On the other side we have small producers that are developing small local niche markets, with an extreme being the "hobby" farmer. Hobby farmers are people that think they like the country life style and buy a few hectares and have one or two cows or sheep or goats or horses. The information needs of these last two groups may be a lot more basic than for the large commercial farms. How do our Extension personnel handle this range?

The changing demographics in the United States are also changing our political reality. The majority of our voters, and therefore our legislatures, are urban. Many have little or no relationship with Extension and therefore can't be expected to support Extension funding.

The other major issue for Extension is the increasing polarization of our public over contentious issues. These involve animal rights and various environmental concerns related to animal production. We are increasingly being asked to take sides instead of being an impartial source of scientific information (Figure 3). The parties on the various sides of the argument question whose science is better. In other words, if the science doesn't agree with their bias, it is automatically deemed invalid. The Extension faculty are spending much of their time dealing with environmental issues and in many cases getting caught in the middle of very polarized arguments. Faculty members are trained in the disciplines of nutrition, genetics, physiology, or range management; therefore, they are being asked for expertise that is often outside their comfort and interest area. Because of these problems in building consensus, issues mediation has also become a need for our Extension professionals and is an area that will certainly increase in the future.

INCREASINGLY ASKED TO TAKE SIDES!

- Environmental and other issues in United States becoming very polarized
- My science is right your science is wrong!
- University and Extension asked not be neutral and just present facts, but to be on one side or the other
- Extension gets caught in the middle

Figure 3. The major challenge facing extension in Oregon.

How do we see Extension changing?

In Animal Sciences at Oregon State University we feel that all faculty have an outreach role. Oregon State University has adopted a system where Extension field faculty have their tenure in academic Departments. Field faculty are those persons who have an Extension or Experiment Station (off campus research center) appointment and are not located on the main campus. Extension personnel that are off campus may have a local or regional assignment. For instance, the Animal Sciences Department has 8 faculty members with Extension appointments in Oregon Counties. I feel that this system strengthens the technical competency of our Extension faculty. The Oregon system also catalyzes a free flow of information in both directions between local producers and the University.

The Animal Sciences Department believes that all Extension professionals must be technically competent. They need some specialization and I personally feel that they should all have some research expectation with their position. Maintaining an active research involvement will keep our faculty progressive. The Oregon tenure system requires that all faculty maintain some form of scholarly activity which also helps to keep our Extension faculty very progressive.

Web based information is continually reducing the need for the more generalist approach to handling Extension education. It may also mean an Oregon producer is getting answers to their questions from Oklahoma State University or Ohio State University instead of Oregon State University. As a result we feel that our Extension programming needs to be directed at very targeted programs. Examples are Quality Assurance (described above), winter nutrition programs for beef cattle and animal waste management programs.

Our larger producers that generate their primary incomes from animal agriculture are increasingly utilizing private consultants as advisors for their production questions. As a result of the greater availability of information and the changing demographics of agriculture described above, our Extension service has matured and is increasingly being asked to train the consultants and other advisors that our producers use.

It is imperative in the Oregon system in the United States of America that Extension be viewed as technically competent, scientifically based and neutral to the issues. With budget constraints reducing the number of personnel we have available more and more of our Extension programming will be regional and state wide in nature and less and less will be one on one with producers. Extending knowledge to producers will, however, continue to be an important part of the three missions of our Land Grant University system in the United States of America.

Extension work in milk and beef production in Slovenia

Marija Klopcic and Joze Osterc

Biotechnical Faculty, Zootechnical Department, Groblje 3, 1230 Domzale, Slovenia

Summary

In CEE countries a trend of decreased milk and beef production has been observed after the year 1990. In Slovenia, milk deliveries increased 30 % in this period, while veal and beef production stayed at the same level as it was in 1990. Milk yield of recorded cows increased from 4,131 kg in 1990 to 5,561 kg in 2002, and the percentage of recorded cows from 33 to 66 % of all dairy cows. Both, microbiological and hygienic quality of milk substantially improved. In 1994, only 60 % of purchased milk contained less than 100,000 m.o./ml, and 97 % in 2002. In the same period the percentage of milk with less than 400,000 SCC/ml increased from 75 to 93 %. The improvements are mostly due to efficient extension work, backed by the Zootechnical Department of Biotechnical Faculty. Up to the year 1990, the extension work was performed through farm cooperatives, partly financed by the state. After the independence of Slovenia the extension work became part of the Ministry of Agriculture and was financed from the state budget. Since 2000, extension work as well as milk recording and animal breeding have been under the responsibility of the Agricultural and Forestry Chamber. The government finances most of these activities. These changes ensured the continuation of successful extension work. Recently, a change of A4 to alternating (AT4) recording method is proposed. Concerning quality, further activities are under way to improve research, educational and extension work.

Keywords: milk and beef production, improvements, applied research work, extension work

Introduction

Cattle production is the most important agricultural activity in Slovenia. Slovene farmers get 40 % income from produced milk and meat. It has been like this since the second half of the 20th century. Cattle production has always contributed a significant part of animal proteins in people's diet. In Slovenia cattle production is specially important also because two thirds of agricultural land is covered with grass. It is in most cases cattle which helps to keep our land cultivated. Therefore it is easy to understand, that cattle production has always gained the special attention of experts and the current agricultural policy. These are the main reasons why in Slovenia, even before its independence in the year 1991, the expert extension service has been well developed. Also applied research work has been very interested in solving the production problems on our farms. To understand the recent cattle production and development it is important to get acquainted with the work of our extension services. The second important segment is to know the ways of our scientific work – research organizations, especially of Biotechnical Faculty, and their part in solving today's problems, as well as creating plans for future development and starting of new programmes. Because of the successful extension services and its cooperation with research organizations it is possible to explain the production quality and quantity in Slovenia in comparison with other CEEC countries.

The history and organization of extension services required for the needs of cattle production

The extension services in cattle production started towards the end of 19th century, when teachers of cattle husbandry were introduced and employed by the Austro-Hungarian monarchy. Between the World War I and II, development in this direction almost stopped. After World War II, the government abolished the agricultural co-operatives of the Soviet type and started with classical co-operative work in 1952 and 1953, which offered some experts to give advice to farmers. Their activities continued until the year 1960, when the agricultural co-operative work ended. After that came 5 years of forced socialization of Slovene agriculture. The politicians were absolutely sure that the big large state farms (large farming enterprises) would produce enough food for everyone and keep the land cultivated. But the deficit of agricultural products was getting worse and in connection with the economic reform in the year 1965 a new idea was born. The new idea was, that co-operating farmer produced agricultural products for large enterprises through contract (large state farms and social type of cooperatives), which took over marketing of the products. So, the new type of socialistic co-operative was formed, with their own production, and teams of expert extension service. This is important, because the extension workers covered first of all their own needs, but soon they started to give advice also to the co-operative farmers. Such cooperation increased the production. As already mentioned, co-operatives organized the market for the entire agricultural production. The law on co-operative work in 1972 insured co-operative relationships and the way of production, which introduced a farmer as a client. In this law it was clearly written, the co-operative can employ an expert (agricultural extension service) to give the farmer advice in production problems. These experts were paid with the money from the sold different material to farmers and sold agricultural products. Because of that law, the co-operative union started working again and in 1975 the union established the **Centre to promote agriculture**. This centre joined agricultural advisers in co-operative teams. Experts in the Centre were financed from the state budget. In February 1982 the important **Law on food production** was accepted. That law provided 10 % co-financing of 190 co-operative advisers, and 40 % co-financing of 30 experts –specialists in district institutions. In 1973 the Animal Production and Business Association, founded by co-operatives, large state farms and the food processing industry gave other funds to pay the specialists. The Association collected funds for its operation, to run its professional activities and control service, special applied research work and for co-financing the extension service, which had as many as 450 advisers in 1986. They were 1/3 financed by the state, 1/3 from the co-operative funds, and 1/3 by the communities. The advisers worked at co-operatives and their work was closely connected with the work of specialists at district institutions. They covered all the needs in the field of agriculture. Because of the fact, that the Slovene farmers still get 40% of their income from cattle production, we can understand, that their work was orientated to advising also on feed production, and improved milk and beef production.

In 1990 great changes took place. The Animal Production and Business Association stopped operating, after that the state of Yugoslavia fell apart, and in 1991 Slovenia became independent. There was a possibility, that the organized extension service would end, as well as its work and that would be a major damage for the Slovene economy. To avoid this, the minister of agriculture moved the extension service under the frame of the Ministry of agriculture, forestry and food (MAFF) in 1990. Two years later milk recording and selection services were given into the hands of the ministry. The number of advisors decreased to 300. Because of the Law on Agriculture and Forestry Chamber (AFC) in 1999, the extension service could become the chamber's responsibility. It happened in October, 2000. The chamber provided financial means, so the extension service was co-funded up to 70 %. Today, 323 advisors are working in eight district institutions which are part of the Agriculture and Forestry Chamber. 92 extension workers are specialists and 18 of them are responsible for cattle production.

Table 1. The History of Extension Service.

Time	Employer	Financing by
End of 19th cent	State / instructor	State
1952 – 1960	Cooperatives	Cooperatives
1965 – 1972	Large state farms + Cooperatives	Large state farms + Cooperatives
1972 – 1982	Cooperatives	State + Cooperatives
1982 – 1990	Cooperatives + district institutions	State (1/3) + Cooperatives (1/3) + Communities (1/3)
1990 – 2000	State (MAFF)	State (MAFF)
Since 2000	Agricultural and Forestry Chamber	MAFF (70 %) + AFC (30 %)

The Agriculture and Forestry Chamber also took over milk recording and selection services and got 80 % financial support from the state budget. 259 recording persons are responsible for milk recording and selection services, 27 of them work in the republic institutions.

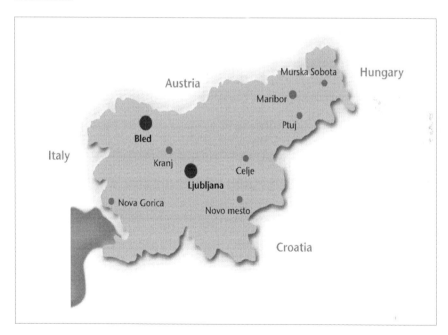

Figure 1. Eight district institutions.

Until recently, the extension service, which was created to help farmers, has been supported by the state and other institutions. Farmers have always regarded extension services as state services and in most cases their attitude has not changed. Somehow they feel, that it is the duty of the state to help them with advisors. Most probably this philosophy is the seed of the past, which will not be easy to stop. Most farmers do not even think about trying to organize and take professional and material responsibility for their own decisions. The change in their attitude can not be expected from the people who have farming just as a supplemental activity, but we can expect a different attitude from the professional, full-time farmers, whose only income comes from the agricultural activities. The same way of thinking is apparent, when we talk about financing the recording service, herd book and selection. However, we can expect, that the professional farms, which will be created by the improved structural changes in agriculture, will soon realize the beneficial meaning of this extension service and will be prepared to pay for it.

Milk production improvement

The amount of milk production depends upon the total number of cows in country (both suckler and dairy), the number of dairy cows, and the milk yield per dairy cow. In Slovenia there are altogether about 190,000 cows, of this 60 % are dairy cows and 40 % suckler cows. For some decades, artificial insemination has been wide spread. Over 90 % cows are inseminated. From the number of inseminations in consecutive years, we can see the movements in the number of cows in the country. The information in Table 2 can confirm, that from 1985 to 1990 the number of cows mostly decreased (over 10 %). The downfall in the last 12 years came only to 6 %. The decrease is even lower if we consider natural mating in suckler cow production on the pasture.

Table 2. Changing number of inseminations.

Year	SI*	B	BW	CH	LM	BPBG	RC	Together
1976	109,629	71,092	11,695	1,334	5,762	-	1,095	200,607
1980	113,077	63,198	16,097	8,191	5,518	-	419	206,968
1985	126,521	73,505	20,103	4,700	2,164	-	160	227,162
1990	116,642	56,262	22,672	3,955	3,481	-	42	203,054
1995	119,260	45,079	25,468	4,650	5,784	123	170	200,534
2000	113,827	29,338	33,257	2,689	11,564	6,432	359	197,484
2001	112,161	27,682	35,410	2,638	12,727	7,275	417	198,300
2002	107,764	24,849	36,409	2,493	12,703	7,663	465	192,346
2003	105,512	22,130	34,784	2,426	13,374	7,714	558	186,498

*SI – Simmental breed, B – Brown breed, BW – Holstein-Friesian breed, CH – Charolais breed, LM – Limousin breed, BPBG – Belgian blue breed, RC – Red Cika Breed

In the last decades, the number of dairy cows included in milk recording has risen (Table 3). On farms, which were until 1991 state owned estates, the number of cows has been falling, but on the other hand, the number of cows on family farms has increased. Leading dairy farms and professional farmers value the information provided by milk recording services more and more, because milk recording results help them in selection and farm management. Farmers gain knowledge from the experts on selection and extension service, through different forms of permanent education, and from the media. Farmers, who realised the benefits wish to be included in milk recording process. Because of this, the number of recorded cows is rising every year (Osterc et al., 2002).

Table 3. The number of cows included in milk recording per sector.

Year	Farming Enterprises	Family Farms	Total
1959/60	-	-	10,478
1966	14,238	7,251	21,489
1972	10,393	25,318	35,711
1980	9,993	54,634	64,627
1985	9,747	49,147	58,894
1990	8,061	50,063	58,124
1995	6,808	59,404	66,212
2000	5,421	65,709	71,130
2002	4,812	69,188	74,000

After the year 1990, milk yield of milk recorded cows increased very rapidly (Table 4). The average milk yield of recorded cows increased 122 kg/year. Milk fat also increased 0.44% and protein contents were up to 0.26 %. The highest progress was observed in milk production of Black and White cows, more than 1,400 kg. The least progress was observed in milk yield of dual-purpose cows, not more than 1,170 kg (Table 5). However, the value of milk fat in dual-purpose cows increase for over 0.5 %.

Table 4. Average milk yield in standard lactation of recorded cows from 1980 – 2003.

Year	No. of rec. herds	No. of rec. cows	No. of finished lact.	Production in 305 days		
				Milk, kg	Fat, %	Proteins,%
1980		37,757	32,418	3,982	3.76	-
1985		58,894	55,873	3,596	3.73	-
1990		58,124	50,994	4,092	3.74	-
1993		63,316	53,290	4,136	3.84	3.08
1995	7,828	62,560	55,450	4,505	3.94	3.19
1997	7,385	70,516	64,701	4,615	4.06	3.24
2000	6,227	67,838	55,603	5,240	4.12	3.34
2001	6,127	69,535	57,589	5,452	4.14	3.34
2002	5.850	74,000	64,999	5,561	4.18	3.34
2003	5,789	75.790	64,426	5,601	4.15	3.33

Because the number and the share of cows, included in market milk production is falling (Table 6), the rate of recorded cows increased to 65 % (comparison of data in Tables 3 and 6). The desire to include new breeds in production is still present among breeders. This proves they know about the benefit brought by production recording. Structural changes in the last 12 years (Tables 6 and 7) are very pleasing. The number of farms included in selling milk fell by over 2/3, but the quantity of sold milk to dairies increased over 30 %. This situation is possible because of the high increase in milk yield of cows. The quantity of sold milk per cow has almost doubled, it was over 4,600 kg and over 46,000 kg per farm in the year 2004.

Table 5. Average milk yield of recorded cows on farms in different years and breeds in standard lactation.

Year	Simmental breed				Brown breed				Black/White breed			
	No. of cows	Milk Kg	Fat %	Prot %	No. of cows	Milk kg	Fat %	Prot.%	No. of cows	Milk kg	Fat %	Prot %
1970	3,857	3,563	3.79	-	7,000	3,386	3.78	-	3,017	4,010	3.79	-
1975	5,151	3,372	3.75	-	7,180	3,513	3.76	-	4,825	4,359	3.69	-
1980	13,968	3,668	3.81	-	9,880	3,744	3.73	-	7,560	4,862	3.73	-
1985	26,539	3,185	3.77	-	16,753	3,513	3.71	-	10,768	4,705	3.65	-
1990	23,674	3,518	3.74	-	14,285	3,902	3.80	-	11,623	5,489	3.66	-
1993	26,239	3,522	3.85	3.18	13,799	4,043	3.89	3.10	12,306	5,543	3.79	3.01
1995	26,092	3,837	3.94	3.24	14,037	4,288	3.98	3.19	14,358	5,930	3.92	3.14
1997	30,327	3,951	4.06	3.26	16,872	4,446	4.08	3.25	16,395	6,019	4.04	3.20
2000	24,327	4,405	4.17	3.38	13,001	4,979	4.15	3.36	17,164	6,633	4.05	3.28
2001	24,747	4,588	4.22	3.39	13,109	5,118	4.16	3.38	18,484	6,860	4.07	3.28
2002	27,168	4,689	4.26	3.39	14,322	5,161	4.19	3.37	21,970	6,914	4.11	3.28
2003	27,130	4,772	4.23	3.38	13,574	5,181	4.16	3.37	22,014	6.858	4.09	3.26

Table 6. The number of herds and cows included in milk purchase and the quantity of sold milk.

Year	Number of herds	Number of cows	Sold milk in litres to dairy industry			Milk content, %	
			Milk, (total)	Per cow	Per herds	Fat	Protein
1980	55,533	150,694	303,831,000	2,016	5,471		
1985	58,194	175,696	352,454,200	2,120	6,063		
1990	43,656	161,992	359,184,200	2,217	8,228	3.74	-
1993	36,327	148,802	346,095,000	2,326	9,527	3.78	-
1995	30,040	132,532	388,394,400	2,968	12,942	3.92	3.24
1998	21,373	122,749	420,127,700	3,269	19,657	4.08	3.33
2000	16,869	117,775	447,831,000	3,758	26,516	4.10	3.36
2001	13,360	116,000	460,562,960	3,970	34,473	4.12	3.34
2002	12,274	114,000	473.500,000	4,154	38,577	4.13	3.33
2003	11,000	110,000	484.179,900	4,402	44,016	4.14	3.34
2004	10,500	105,000	486.009,740	4,629	46,287	4.18	3.36

Data from Statistical Gazette (1980, 1985, 1990, 1993, 1996, 2000, 2001,2002, 2004) and Internal Reports GIZ – Animal Production Business Association of Slovenia (1985, 1990, 1993, 1995, 1998, 2000, 2001,2002, 2004).

Table 7. Structure of herds and the number of cows in herds on family farms, included to milk purchase.

Year	Farms and number of cows/farm, %				No. farms	No. cows/farm
	1-4 cows	5-9 cows	10-15 cows	Over 15		
1981	78,2	19,2	2,1	0,4	52.221	2,78
1985	78,6	18,0	2,7	0,7	58.130	2,86
1990	73,5	21,3	3,6	1,6	43.613	3,53
1992	71,8	22,0	4,6	1,6	38.154	3,92
1995	62,0	28,6	6,7	2,6	30.012	4,36
1997	58,5	28,1	9,4	4,0	25.063	4,90
2000	46,9	30,0	13,7	8,9	16.847	6,79
2001	41	32	17	10	13,360	8,68
2002	36	34	18	12	12,274	9,29

Knowledge transfer in cattle husbandry

The average number of cows per farm has almost tripled, from 3.5 up to 9.3 cows per farm in year 2002 (Table 7). The number of farms with just a few cows has reduced. The so called half-farms, which get their income also outside the agricultural activities will stop milk production, as soon as their income from other than agriculture activities will insure them a stable standard.

Very important is the fact, that in the last few years breeders greatly improved also the microbiological and hygienic quality of milk (Tables 8 and 9). In the year 2002 only 2-3 % of sold milk did not meet the European requirements. Less than 8 % of sold milk had over 400,000 somatic cells. The improvement was achieved due to the penalties and rewards policy, imposed by the state (Ministry of Agriculture, Forestry and Food) in 1993. Better milk quality was recorded. Such policy was possible in times, when the state dictated milk prices. But for sure, breeders will need to do hard work to improve milk quality and to keep it on the high level through the year. In some summer months, the quality is a bit lower (Figure 2). Milk quality and content are getting more and more important and in the near future, the composition and structure of milk proteins will play a major role (Osterc & Klopcic, 2003).

Table 8. Microbiological quality of sold milk to dairies.

Year	Sold milk in different quality-classes, % (No. of microorganisms in ml)		
	Up to 400,000	Up to 100,000	Up to 50,000
1994	87.2	60.40	43.50
1995	93.0	78.60	61.49
1996	93.7	75.70	63.31
1997	94.7	82.82	68.33
1998	95.7	84.10	69.21
1999	96.9	85.88	69.86
2000	97.9	95.16	85.71
2001	99.1	96.74	90.55
2002	99.4	97.35	91.21
2003	99.5	98.12	91.81
2004	99.5	98.57	92.94

Table 9. Somatic cell count in sold milk.

Year	Sold milk in different quality-classes, % (Somatic cell count in ml)		
	Up to 400,000	From 400,001 to 600,000	Over 600,000
1996	74.96	20.06	4.98
1997	80.35	16.06	3.59
1998	81.90	15.01	3.09
1999	85.01	12.82	2.17
2000	91.08	7.84	1.08
2001	93.38	5.82	0.80
2002	92.76	6.36	0.88
2003	92.19	6.99	0.82
2004	92.72	6.50	0.80

Recording results (Figure 3) show substantial difference in the somatic cell count (SCC) among breeds. The lowest SCC is found in milk of Simmental cows, the highest in Black/White cows. Due to the growing rate of Black/White breed in the dairy cow population and at the same time the growth of milk yield of the same breed, milk quality regarding SCC does not improve any more (Table 9).

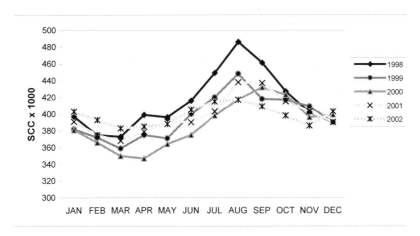

Figure 2. Average somatic cell count – milk recording results.

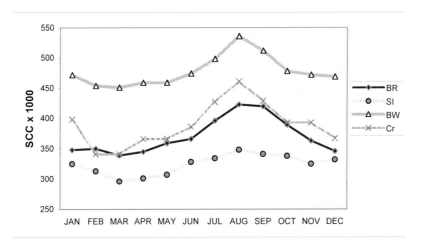

Figure 3. Average somatic cell count per recording month and per breed.

Improvement in beef and veal production

After World War II, there was a great demand for quality beef in Europe. High prices and open market meant a stimulation for cattle breeders. In Slovenia breeders kept mostly dual purpose cattle breeds: Brown and Simmental. Young animals of both breeds were very suitable for fattening, and fattened animals gave quality meat. Slovene breeders were therefore oriented to fattening of young cattle for the market. Data show that sold meat brought them much higher profit than sold milk (Table 10). Up to 1985 the income for sold meat in comparison to sold milk was gradually decreasing and in 1985 it was equal. After the year 1985 the income from sold milk increased. In the last 5 years the proportion has been largely unchanged (Cepon *et al.*, 1998).

Table 10. Changing the value of purchased cattle and milk in millions din up to 1991 and millions SIT after 1991.

Year	Animals (slaughter of cattle and calves)	Milk and dairy production	Proportion (meat : milk)
1902	8,3 mio K*	1,1 mio K	7.5 : 1
1956	4,8	1,4	3.4 : 1
1960	76,7	25,6	3.0 : 1
1965	330,6	78,5	4.2 : 1
1970	414	144	2.9 : 1
1975	1.009	599	1.7 : 1
1980	2.408	1.972	1.2 : 1
1985	14.578	14.305	1 : 1
1990	764	1.256	1 : 1.6
1995	8.703	15.446	1 : 1.8
2000	10.932	26.047	1 : 2.4
2001	11.910	27.050	1 : 2.3

Source: Statistical yearbook of Slovenia;   *in crowns

Table11. Cattle fattened and slaughtered in slaughter houses in Slovenia.

	Number in 1000		Carcass weight in 1000 tons	Estimation of body weight in 1000 tons
Year	Slovene only	Total slaughter	Slovene only	In Slovenia
1955	173	173	19	38
1960	159	159	18	56
1965	187	187	32	75
1970	138	180	26	60
1975	181	221	37	79
1980	136	152	36	85
1985	144	177	39	86
1990	147	172	39	88
1995	111	120	33	88
2000	126	130	34	91
2002	154	157	43	98

Source: Statistical yearbook of Slovenia and calculations based on the information in a yearbook

Until the year 1985 a growing trend in the number of beef cattle is observed. In the years of Slovenian independence, production has slightly decreased, but in the last few years it has, once again, risen over 10 %. After the independence of Slovenia, the purchase of slaughter animals also decreased, specially because it was rather high in the time of the former Yugoslavia (Table 11). In these years, the proportion of slaughter categories changed, too. The share of veal at around 20 %, is still relatively high, but the largest part is represented by young fattened cattle. Lately, this rate has fallen due to the increased number of cows (Table 12), especially Black/White cows, which are very popular by milk producers. Black/White cows stay in production for a shorter period, and breeders have to replace a larger percentage every year (Osterc *et al.*, 2001) .

Table12. Structure of slaughter in Slovene slaughter-houses (in 1000s).

Year	Calves		Bulls	Cows		Oxen	Young cattle	
	Number	%	Number	Number	%	Number	Number	%
1955	110	64	4	29	17	16	13	7
1960	69	43	19	31	19	19	20	13
1965	57	30	35	26	14	12	57	30
1970	59	33	21	28	16	13	59	33
1975	69	30	31	19	9	5	97	45
1980	19	12	-	20	13	1	112	73
1985	14	8	-	20*	11	-	143	81
1990	13	7	-	19	11	-	142	82
1995	14	12	-	15	13	-	90	75
2000	29	22	-	20	16	-	79	62
2002	28	18		36	23		92	59

* after the year 1985 the older bulls and oxen are marked in category of cows
Source: Statistical yearbook of Slovenia and calculations based on the information in a yearbook

Recently, changed breed structure (Table 2) and changed types of Simmental and Brown breed have also brought the change in carcass quality of slaughtered cattle (Figures 4 and 5). Carcass conformation is lower, as well as fatness of some slaughtered cattle categories. Figures 4 and 5 are made on the base of data gathered in bigger slaughterhouses referring to 60 % of slaughtered cattle last year, so data can be considered as fairly reliable (Zgur, 2003).

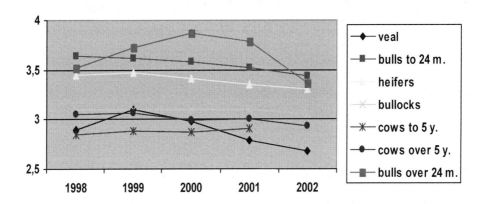

Figure 4. Average score of carcass conformation for different categories (E=5, U=4, R=3, O=2, P=1).

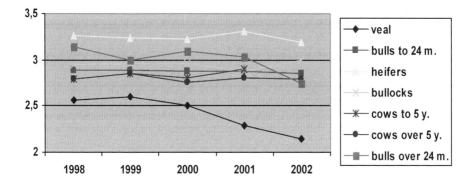

Figure 5. Average score of carcass fatness for different categories.

Applied research work and professional services

Increased milk and meat production, and their quality after the year 1991 is a consequence of many factors. The progress is certainly due to the professional services, especially the selection and extension services. The quality work of these services is based upon good organisation, and the way these services are provided with new knowledge, which they have to forward to their customers, the farmers. The organisation was in time of transition, when there was danger that these services will disappear like in other CEEC countries, provided by the Minister of Agriculture. Till the foundation of Agriculture and Forestry Chamber, he transferred their organisation and activity under the frame of the Ministry for Agriculture, Forestry and Food (MAFF). Experts and researchers required knowledge and it was provided by the Agricultural Institute of Slovenia (AIS), district agricultural units, secondary schools, Faculty of Agriculture Maribor, and in the last two decades particularly by the Biotechnical Faculty, Zootechnical Department. Along the financial support the Ministry for Agriculture, Forestry and Food (MAFF) dedicated most of the time to the applied developmental research and to the organization of different forms of permanent education (Osterc & Fercej, 2000).

The whole time farmers were advised by experts, their knowledge was based on local research results that have also considered special Slovene cattle breeding conditions. Especially important was the research work and pedagogical work at the Biotechnical Faculty, Zootechnical Department. They had a special task to investigate how to increase production, to investigate supplementary methods in data processing, and the supplementary breeding value calculations. The results of these researches provide the base for the improvement of breeding and production programmes, which promote better ways of using production potential of cattle in Slovenia. Among the important research that were going on in the past are the possibilities and the improvement of suckler cow production, crossbreeding of beef cattle with our dual-purpose breeds, investigation of industrial crossing to acquire quality calves for fattening to higher weight, and also to investigate the most suitable breed combination for suckler cow production, introduction of linear cattle evaluation and the utilisation of the achieved results for management purposes, introduction of BLUP and ANIMAL models for data processing and breeding value calculation of separate traits for the needs of selection in cattle production; development of reference A4 milk recording method with monthly prints and including new milk traits to the recoding process.

The results of all these researches are available for demonstration experiments, where professional service workers are involved, as they are involved in other kinds of permanent education. Advisors are qualified to put new knowledge into farming practice.

Introduction of new programmes

Certainly, the application of modern methods in milk recording and the activity of advisors is never concluded. That is why the experts at the Biotechnical Faculty, Zootechnical Department are developing new milk recording programmes and the programmes for the needs of extension services in cattle production, to offer help in the restructuring processes, and to supplement the existing programmes.

Momentarily, the most urgent question is cheaper milk recording, which would increase the number of information for breeders, but reduce milk recording costs at the same time. The possibility to change from reference method A4 to method AT4 was thoroughly investigated. For Slovene circumstances correctional factors were calculated by modern computers for the calculation of standard lactation, that will in spite of cheaper recording remain equally reliable as before. At the beginning of the year 2004 we started with the introduction of AT milk recording method as everyday practice (Klopcic, 2004).

Monitoring of the increased beef and veal quality (Figure 4 and 5) has shown, that the higher rate of Black/White breeds and the changed combining of Brown type, partly also Simmental type, worsened the quality of slaughtered cattle. Breeders are trying to overcome this worsening by industrial crossing with beef breeds (Table 2). The efficiency of industrial crossing also depends upon the decision which beef breed to choose for crossbreeding. Now, extensive experiments are in progress, investigating crossbreeding with Charolais, Limousin, Blond'Acquiten and White/Blue Belgian breed. We expect to establish the most suitable crossing combination for Slovene conditions to improve the quality of fattened cattle once again.

Customers demand beef and veal from different sustainable forms of cattle production. That is why veal production in the form of suckler cow production is intensely examined. In Slovenia there is a variety of conditions for suckler cow production, therefore it is not possible to recommend only one breed or one type of cows for this intention. We hope that experts from the Zootechnical Department at the Biotechnical Faculty will give answers to this question, and advise which suckler cow breed should be most suitably used in different environmental conditions. This project is extremely important, namely because it is estimated, that Slovene breeders are already rearing more than 1/3 cows as suckler cows. For this kind of production the farmers get financial state premium. The share of suckler cows will as yet rise with structural changes.

Because Slovenia is too small to have a separate institution just for applied research, we organized this kind of research work at the Biotechnical Faculty, Zootechnical Department decades ago as university graduate and postgraduate studies for the needs of cattle production. Thus, students get acquainted with the research problems during the course of their study. So after graduation they work as advisors or experts in recording and selection services. They are best qualified to take part in developmental applied research, in the transfer of new accomplishments to farmers, and to participate in different forms of permanent education, which are also organized mainly by the already mentioned institution.

Conclusions

Better milk yield, higher milk quality, changes in beef production, as well as great structural changes in the last decade are mainly the consequence of numerous factors. In comparison with the changes in other CEEC countries, also some EU countries like Austria, Greece, Portugal, we can be proud of the achieved results. It is true that the Slovene cattle breeders are gradually more educated and respect new knowledge, but without the efficient work and cooperation with advisors, recording persons, and experts at the educational-developmental institutions, the success could not have been achieved. The results clearly indicate that the extension services have improved their organisation, and because they were not cancelled in the time of transition, as happened in other CEEC countries, the positive situation in Slovenia developed. The organisation and function was always adapted to Slovene conditions and needs. However, specialised dairy farms have to intensify and devote their efforts to the elimination of deficiencies in farm management, especially after the introduction of milk quotas. We are sure they will see and feel the solution for their problems in adequately qualified extension workers. Therefore the educational and research institutions in Slovenia are faced with an important task – to educate specialists, which will be capable of helping the farmers. Here, foreign experience will be respected, yet adapted to specific cattle production conditions in Slovenia.

Figure 6. Applied knowledge about mountain pastures also of importance.

References

Cepon, M., S. Cepin, & C. Varga, 1998. Quality of animal production and animal products in Slovenia. Research Reports Biotechnical Faculty University of Ljubljana. Agriculture. Supplement 30, 15-23

Klopcic M. 2004. Optimization of milk recording practices in dairy cows. Dissertation. Biotechnical faculty, Ljubljana, Slovenia, 171pp.

Osterc J. & M. Klopcic, 2003. Pomen kakovosti mleka za proizvajalca. In: Referat na posvetu za SANO, Velenje, Slovenia.

Osterc J., M. Klopcic, K. Potocnik & M. Cepon, 2002. Napredek slovenske govedoreje v zadnjih stirih desetletjih. In: Razstava zivali 2002. 40. mednarodni kmetijsko-zivilski sejem, Gornja Radgona, Slovenia, 10-23.

Osterc J. & J. Fercej, 2000. Razvoj trzne prireje mleka in mesa v Sloveniji. Govedorejski zvonci, 5, 1/2: 7-8

Osterc J. et al., 2001:The future of milk and meat production in Slovenia. Research Reports Biotechnical Faculty University of Ljubljana. Agriculture. Supplement 31: 5-18

Pravilnik o ocenjevanju in razvrscanju govejih trupov in polovic na klavni liniji. 1994. Ur. l. No. 28-26: 1831-1834

Pravilnik o ocenjevanju in razvrscanju govejih trupov in polovic na klavni liniji. 2001. Ur. l. No. 103: 10881-10885

Zgur S., 2003. Kakovost zaklanih goved v Sloveniji od leta 1998 do 2003. Sodobno kmetijstvo, 36: 9

Knowledge transfer in Slovak cattle production during the transformation period

Stefan Mihina[1], Bill Mitchell[2] and Vojtech Brestensky[1]

[1] *Research Institute for Animal Production, Hlohovska 2, 949 92 Nitra, Slovak Republic*
[2] *Extension specialist, 3 Woodlands Grove, Ilkley, LS29, 9 BX, United Kingdom*

Summary

In Slovakia, as in other Central and Eastern European countries, major changes have occurred in all areas of life. The conditions for cattle breeding have changed significantly, mainly in relation to product markets. Domestic consumption has decreased, and the milk and meat markets have been liberalised. Pressure on product quality has increased. Simultaneously, however, the opportunities for technological and biological modernisation have improved. A system of extension services was required which would be widely available. A flexible Extension Services Network was therefore introduced in which all could engage – independent advisers as well as advisory organisations. Advisers as well as users have become familiar with new ways of communication with the aim of utilising the most recent knowledge in practice as effectively as possible. The paper will describe examples of production system approaches to the modernisation of cattle farms in Slovakia. Also the role of the Animal Production Institute in this process of knowledge transfer will be discussed. Some findings of the EU-project, in which the dissemination of research into practice under the conditions of the enlarged European Union were studied, will be presented.

Keywords: cattle production, knowledge transfer, transformation period, extension system

Cattle breeding in Slovakia

Predominantly large-scale farms are used in cattle breeding in Slovakia at present. The size of farms has changed very little compared with the period before 1990. However, the management of farms has changed significantly.

In spite of there being large farms in Slovakia before 1990, there was a high proportion of tying stalls with milking into a bucket or through pipeline milking installations. There was also a high proportion of dual-purpose breeds with quite low performance. The quality, and sometimes also the quantity, of roughage were insufficient.

Immediately after 1990, the market for agricultural products opened up. The short comings in the management of breeding cattle could not cope with this situation, therefore a very rapid and significant decrease in the amount and intensity of breeding took place. Numbers and performance in cattle both decreased a lot. The largest depression was noticed in 1992. However, as a result of the open borders, farmers got new opportunities for biological and technological modernisation of farms.

The government supported modernisation, but a serious problem became apparent. There was no extension system in agriculture at that time. Before 1990, dissemination of knowledge was effected by means of a network of regional offices of the Ministry of Agriculture. The knowledge was not offered, but was ordered (by force) in most cases. In some districts, for instance, it was only possible to keep a certain type of cattle with specified technology, by order of the government and regional office. Rotation of crops was ordered, the areas of crops, mainly cereals, being controlled. This had a very negative impact on production and quality of roughage. In spite of existing knowledge about behaviour of dairy cows in various types of housing, certain types of housing began to be used that were unnatural for the animals and confined them. Livestock numbers exceeded the housing capacity. Therefore many animals were housed in inappropriate buildings. It was surprising that tying stalls for cattle actually became more numerous. The reason for this was that it was easier to manage the farm. One stockman looked after a certain group of animals. He fed them, littered them, removed manure and, with dairy cows, milked them and was also responsible for the control of reproduction.

However, at that time there already existed a small number of quite successful farms that were created mainly on the initiative of breeders themselves who were able to visit modern farms in the USA. Some farms in neighbouring Hungary were also good examples. Modernization of cattle breeding, mainly dairy cows, had already started there in 1975.

Need for extension services

Farmers struggled to adjust to the new market economy. In particular, they were unprepared for international competition that demanded higher quality standards, attractive presentation, continuity of supply, appropriate transport systems, and a degree of producer co-operation and organisation hitherto unknown. In the circumstances, profits were scarce and losses commonplace.

Government came under pressure to give greater financial assistance to agriculture, but overall state funds were severely limited and other sectors of the economy were seen as having higher priority in the years leading up to and following independence from the Czech Republic in 1993.

For significant progress to be made in Slovak agriculture, it was clear that greater efforts would be required to:
a) create awareness of worthwhile innovations that were already in use on farms in other countries,
b) introduce new systems and technologies, and
c) assist farmers in the transformation process.

Creation of the extension system

As mentioned already, there was no extension system for Slovak agriculture at the beginning of the nineties. Regional offices of the Ministry of Agriculture continued with their activities. Some emerging private trade companies provided extension, usually related to their particular products; research and educational institutions began to encompass extension in varying degrees; and private extension specialists began to function. Also, new private extension companies were created, with the aim of meeting the obvious needs within the industry. But paying for advice was a new concept, its value was difficult to grasp, and in any case farmers already had difficulty paying for more essential and tangible goods such as feed and fertiliser. A self-governing organisation, the Slovak Chamber of Agriculture and Food supported extension significantly and its regional office activities gradually replaced those of the government regional offices. The Chamber tried to co-ordinate extension in Slovak agriculture in its own particular way. For purposes of extension, there arose an association of agricultural research institutes, Agroservis, that also contributed to co-ordination and to the creation of a new system of extension in Slovakia.

It became necessary to create a flexible extension network to accommodate the great diversity in status and objectives of the extension institutions and individual extension workers. Such a network was created as a result of the EU PHARE "DESIPAP" project (Development of Extension Services to Improve Primary Agricultural Production) which ran from 1996 to 1999.

This project was preceded by a number of initiatives, some Slovak but mainly foreign. A number of individuals was trained in extension. Some outstanding foreign extension institutions organised these training courses and they tried to recommend the most suitable system of extension for Slovakia at the same time. We had the possibility of adopting various recommendations from representatives of foreign extension institutions. Some of them recommended creation of the same extension system in Slovakia that they had in their own countries. Pessimists said that no overall extension system could be created in Slovakia.

The team of the PHARE programme proposed the creation of an original Slovak extension system and it was accepted by the government. It respected the current situation that had developed in Slovakia, and it proposed creating an Extension Services Network (ESN) that was open, and remains open, to all who are interested in working in agricultural extension. It was important that the proposal also respected the requirement of the government that it should give only partial financial support for extension. In addition to advising the government on the best structure for the network, the team provided detailed and specific training for those who would be involved in extension work, and organised the provision of suitable computing and communications equipment for extension centres.

The network consists of various groups of participants with specific functions:
- Users of extension (farmers and farm managers),
- Advisers,
- Trainers of advisers,
- Collection and dissemination of information,
- Co-ordination of ESN,
- Funding.

The system enables a given subject to be covered in more than one group. All existing institutions and individual advisers were able to register for the system, and it is expected that newly-emerging ones will also enter it. To be registered, advisers must have a specified level of qualifications and experience. Many are in fact part-time, having main careers in commercial, educational or research organisations. The system is open for providers as well as users of extension.

The training centre, Agroinstitute, supported by other existing educational institutes and individual tutors, provides basic training and updating for advisers. Also at the Agroinstitute, there is a co-ordination centre in which all participants in the extension network are registered. It co-ordinates 16 extension centres spread over the territory of Slovakia. This unit can put enquiring farmers and others in touch with the names of the most appropriate advisers for solving their particular problems.

Information is concentrated on a separate website that is gradually filled with information mainly by users of the information. The existing Institute for Scientific and Technical Information in Agriculture administers the ESN website.

The regional offices of the Ministry of Agriculture provide the funding. They give subsidies to users of extension on the basis of the invoice submitted by the provider of extension. A percentage of the proven costs is refunded. Percentages are always set by the government at the beginning of the year. Some preferred spheres of production or regions can be allocated higher percentages, in line with current government priorities.

The ESN is essentially a facilitating organisation. It is not in charge of advisers or other extension organisations, which continue to run their own affairs but use ESN co-ordination services. Staff of the co-ordinating unit do not themselves provide advice to clients and are therefore not in competition with registered providers.

Cattle breeding was one of the spheres in which extension was provided even before the ESN was created. The Research Institute for Animal Production (RIAP) in Nitra was very active in this respect.

Communication between adviser and user of extension

The abovementioned system of dissemination of knowledge by directive influenced the superficial outlook of the early advisers and users of extension. Advisers, most of whom were specialists, were aware of their abilities and were convinced that users should use their recommendations from A to Z. In the beginning, users also expected that they would get 100% proposals for solutions from specialists that they would be able to put into practice without the specialist's presence. However, both parties gradually realised that resolution of a given problem must follow known principles that they adopt step by step. The education of advisers is also helpful in this respect. The following principles illustrate the point:

Formulation of the problem by the farmer himself

The adviser should first of all listen to the farmer carefully and let him formulate the problem by himself. Often it happens that the adviser, focussing of his own abilities, does not pay enough attention to all the problems of the user because they seem to him to be solved already. However, they are important for the user and therefore the adviser should comprehend all the problems of the user, accepting his concerns even though they seem at first sight to be unimportant. It is very important that the adviser knows all the limitations that stop the user achieving an optimal resolution of his problem. If these guidelines are observed, the problem, to be solved jointly by the adviser and the user, should be well defined.

Common search for the most suitable method of solving the problem

The adviser must not exclude the user when looking for a solution. It is more satisfying for the user if he believes that he found the solution largely by himself. The adviser should be able to supply the user with an analysis of the advantages and disadvantages of the various possible methods of solution and support this analysis with technical knowledge. Besides the limitations familiar to the user, there often exist further limitations - external ones - about which a qualified adviser should be able to inform the user.

Adviser should help to reduce the number of possible solutions

Of course, there are many more possible solutions at hand than can be implemented in practice. Therefore it is necessary for the adviser, together with the user, gradually to decrease the number of possible solutions. After they have chosen the most suitable, they should discuss expected results.

To achieve satisfaction of the user

The discussion should end with the user expressing satisfaction. It is not good if the user starts implementation with doubts or with a feeling of dissatisfaction. In such a case, the adviser must suggest to the user that they go back and search together for another solution that will make the user happy. If they reach agreement, it is very important that the adviser participates in the implementation of the chosen solution.

Expectations of the user when implementing the solution

The user should feel that he still needs the adviser during implementation. However, he should not see him as a controller; nor should he have the feeling that the adviser must lead him at every step. Our experience indicates that the results of extension are most effective in circumstances where the farmer is capable of implementing the solution by himself but chooses to have the assistance of the adviser. In this relationship, the adviser should act as a catalyst.

Extension in the Research Institute for Animal Production in Nitra

As mentioned earlier, one of the biggest problems to be solved in Slovak agriculture was that of cattle breeding, mainly in dairy cows. In the Research Institute for Animal Production in Nitra there are specialists in various spheres of animal production. The institute initiated the creation of an association of research institutes for extension in agriculture, called Agroservis.

Since 1993, development projects for advancement in cattle breeding were supported by the state. Mostly, breeders themselves prepared these projects. The RIAP took over the role of adviser at the beginning. During this activity, a need arose to establish guidelines for the methods and principles to be used in the creation of such projects. The processes of project preparation which emerged respect the principles of communication between adviser and user described above.

The institute also elaborated a number of model development projects on real operational farms. A Department for Extension and Marketing was created at the institute. It uses high-calibre specialists in research departments (mainly in the spheres of genetics and selection, management and economy, nutrition of animals, welfare of animals, and buildings and equipment) and sets up flexible teams for individual projects in accordance with the requests of the breeder.

Most projects have been aimed at breeding in dairy cattle. A development project in dairy cattle breeding usually contains:
- An analysis of the herd, highlighting the proportions of breeds, performance, milk composition, number of lactations, and parameters related to reproduction. The analysis is carried out on the basis of data obtained from the breeder and from organisations involved in performance testing.
- Expected future parameters of performance on a time schedule. The projections start with the visions of the breeder, which are then adjusted by the adviser to values that are achievable in practice.
- Larger herds need thorough processing of herd turnover data. Processing is carried out by the adviser, in conjunction with the breeder, in a way that takes into account the current structure and quality of the herd and any possible improvements. Numbers of cattle in separate categories determine housing capacities and the requirements for cultivated feeds. Procedures vary. In many cases, the future size of the herd is determined first, then the capacity of livestock housing, barns and area of fodder crops are adjusted to meet it. In a

number of cases, however, the capacity of buildings and/or the area of crops are limited.

- Stock numbers and planned production determine the total amount of product that is to be placed on the market. There was no milk quota system in Slovakia until this year. Quota limits the current possibilities. Therefore total stock and herd turnover are now, and will continue to be, dependent on the total product that can be placed on the market.

- A suitable selection programme is also determined. Mostly, this involves basic recommendations and identifying present shortcomings that should be improved by implementation of the breeding programme. Actual implementation is carried out by an authorised commercial insemination company, with which the breeder has entered into a contract.

- The programme of modernisation of buildings is a very important element. It starts first of all from the above-mentioned target numbers in the separate categories of cattle and the buildings that are currently available. Livestock decreased by more than 50 % in Slovakia after 1990. Therefore there are usually enough buildings available, and it is mostly older ones that are modernised. However, if necessary, the construction of new buildings is recommended. RIAP occupies an important position in these developments, mainly because it is the author of recommendations that were used in the period of legislative vacuum on animal welfare before EU legislation was officially adopted. It is worth pointing out that the recommended parameters used then are not at variance with the present legislation, so the model projects can still be used.

- Technological modernisation was very demanding in many old building. Problems associated with limited internal dimensions and distances between supporting structures necessitated quite demanding solutions. The institute published detailed modernisation plans in a booklet that was put at the disposal of all breeders and advisers through the extension services network. As a result, acceptable planning methods came into general use.

- The necessary investment costs are calculated from the specific technological plans generated; management of the breeding process is organised and the optimum number of staff is determine. As employees are involved, it is necessary to observe the Labour Code, in which the number of working hours and requisite breaks are fixed.

- The balance and production of feeds are critical parts of the project. The latest knowledge on utilisation of nutrients obtained in the RIAP is used. The institute is administrator of the National Feed Data Bank. These data can also be used in any given project. It is necessary to co-operate with the farmer himself during the preparation of feeding plans, as in other spheres. Natural and technical conditions on the farm must be known in order not to cause environmental, technical or economic imbalance by the intensification proposals. Suitable crop rotations for the production of roughage are elaborated, or pasture and optimal alternatives of feed rations for separate categories of cattle are prepared. Proceeding by stages, the increases in performance that were given in the programme of development are achieved. If the farmer asks for more alternatives for economic or environmental reasons, the specialist checks and quantifies these alternatives.

- Previous recommendations are economically re-evaluated so that the farmer can make up his mind to continue implementation of the programme or to make necessary adjustments.

- The programme is presented in written form, with annexes to supplement it. The annexes include mostly layout of the farm and buildings but there can also be a complete budget for modernisation, a list of proposed sires, alternative feed additives, various prospectuses, etc.

Of course, the process of transformation continues and changes are occurring on many fronts, and in many institutions.

The Research Institute for Animal Production is first of all a research institution and it does not have ambitions to be involved in extension activities across the whole spectrum of animal production. However, research scientists must try to ensure that the knowledge they generate is transferred into practice. The creation of model projects for the development of dairy cattle enterprises is one of many complex examples. The extension activity of the institute starts from scientific results obtained from research, not only in the RIAP but also from across the world, and the identified needs of practical farming in the particular conditions of Slovakia. Information about needs is gained during the 10 – 12 conferences and workshops that are organised for practice each year by the institute, and from many other sources.

At the present time, for certain advanced scientific developments, only the specialist research scientist is capable of transferring the new knowledge on to the farm. In many other areas, a good livestock adviser is well able to absorb the information and pass it on, modified if necessary according to particular farm circumstances. But general advisers, capable of bringing together the various technologies of the whole farm, e.g. livestock, cropping, machinery, labour and finance, are very scarce in Slovakia.

This varied background to dissemination of results creates a dilemma for the RIAP, and also for other research institutes. How far should the RIAP go in terms of dissemination? It already employs a wide range of dissemination techniques, including leaflets, booklets, selected books, web pages, lectures and presentations, on-farm demonstrations, newspaper and magazine articles, scientific journal papers, occasional broadcasts, and training courses. Offering a full range of extension services in livestock production is simply not possible, for various reasons. On the other hand, managers at research institutes no not want valuable new knowledge to wither and die because there are inadequate resources to transfer it on to farms. And a small amount of on-farm extension experience is a desirable component in applied scientific pursuits.

With that problem in mind, the RIAP recently commissioned a study and report covering its present methods of dissemination and suggestions for improvement. The study is complete and the report has just been presented. The report, although in many ways complimentary to the institute, contains a range of recommendations, and these will have to be studied in detail. They cover such broad areas as:
a) creating closer links with practice,
b) using farm systems as the template against which much of livestock research planning and dissemination of results should be seen,
c) the extent to which RIAP scientists should engage in direct extension, and
d) the need for agreed internal policies for both extension and dissemination.

It is always necessary to move with the times, and managers of the RIAP are confident that some of the recommendations in the report, when implemented, will bring further improvements in the work of the institute and in the transfer of new knowledge.

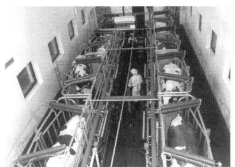

Figure 1. Developments in milking parlours and techniques.

Figure 2. An example of extension efforts of the Research Institute for Animal Production.

Know-how transfer in animal breeding - the power of integrated cow data bases for farmer´s selection of bulls to improve functional traits in dairy cows

Jan Philipsson[1], Jan-Åke Eriksson[2] and Hans Stålhammar[3]

[1] *Dept. of Animal Breeding and Genetics, SLU, Box 7023, SE-750 07 Uppsala, Sweden*
[2] *Swedish Dairies Association, Hållsta, SE-631 84 Eskilstuna, Sweden*
[3] *Svensk Avel, Örnsro, SE-532 94 Skara, Sweden*

Summary

Animal breeding research is usually based on different types of data, e.g. from laboratory analyses, selection experiments and field records. The latter type plays an important role in linking research with practical animal husbandry and breeding. Information about large animal populations in normal production systems is collected from the routine recording schemes, such as milk- and beef-recording, and artificial insemination services (AI). These data provide two essential roles in context of research and know-how transfer: i) In providing data for analyses, research results will be implemented through improved methods directly applied in the recording schemes, or information that these provide to farmers, e.g. improved estimates of breeding values of bulls and cows. ii) Integrating data from different sources enables more holistic farm and animal analyses, and as regards genetics, to estimate breeding values for a range of production and functional traits. In Sweden this integration of different sources of data and computer systems started already in the sixties and has allowed dairy farmers in the last two decades to select AI bulls according to production, female fertility, calving performance, health and conformation traits, all incorporated into a Total Merit Index.

Keywords: cattle breeding, milk recording data, functional traits, integrated data base

Introduction

Animal breeding in general, and dairy cattle breeding in particular, has shown what can be achieved from collective efforts of farmers in applying modern breeding programs. The basics of genetic improvement programs include both genetic aspects of the traits considered and organisational aspects including collaboration among those participating in the breeding program. The integration of genetic and organisational aspects may be illustrated by the different responsibilities that various parties, such as producers, breed societies, AI-organisations, researchers etc, may play in conducting and developing a breeding program (Figure 1). Various parties may play quite different roles, depending on country, tradition, resources and stage of development.

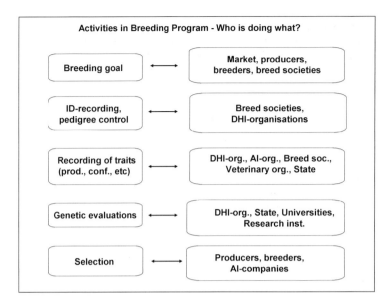

Figure 1. Activities in breeding program.

A breeding program could also be seen as a continuous development project aiming at achieving the goals defined by the farmers. Such a development is consistently dependent on research and application of relevant results. The efficiency of such a knowledge transfer is dependent on many factors including relevance of research as well as efficiency in data flow, extension work and farmer´s application of results as end-users. Figure 2 illustrates very clearly the principal relationship between recording schemes delivering relevant animal data and the continuous feed-back from research for improvement of such practical applications as genetic evaluations and use of the results for mating and selection.

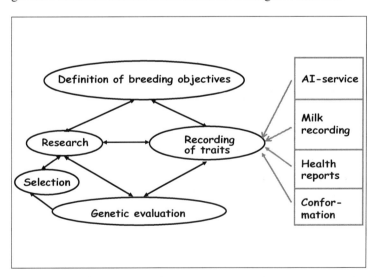

Figure 2. Interactive parts of a breeding programme.

Knowledge transfer in cattle husbandry

As the integration of organisations involved in dairy cattle production and the computerisation of data from milk-recording and other sources started quite early in Sweden it may have some merit in discussing the experiences and philosophies behind this participatory development and the effectiveness in knowledge-transfer as measured by farmer participation rates and actual genetic progress achieved.

The Swedish dairy herd population structure and milk-recording development

The present Swedish dairy cattle population consists of about 410, 000 cows, divided on the two breeds Swedish Red and White (SRB) and Swedish Holstein (SLB) with about equal number of cows. The number of recorded cows amount to 86 % of the total cow population. From Table 1 and Figure 3 it can be seen that in 1960 only 25 % of the 1.1 million cows participated in milk-recording.

Table 1. Number of dairy herds, cows, and average production in milk-recording 1960-2000.

Year	% of cows in milk recording	Number of cows in milk recording	Number of recorded herds	Milk yield kg FCM
1960	25	283,330	23,724	4,455
1970	49	353,146	24,320	5,209
1980	64	410,480	17,982	6,044
1990	75	370,067	14,891	7,319
2000	86	355,197	9,115	8,612

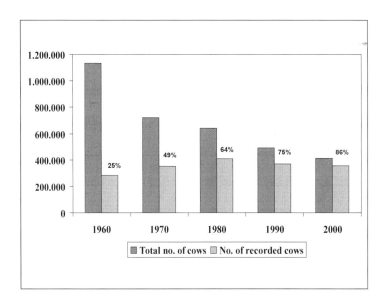

Figure 3. Number of Swedish dairy cows.

In 1960 all records were processed by special milk-recorders visiting the farms. Due to the high costs and the technical development the Swedish Livestock Improvement organisation

(SHS) introduced a computerized milk-recording scheme. Soon after that a milk-recording scheme based on farmer´s own weighing and sampling of the milk was introduced as the official recording scheme, as opposed to what was considered official in most other countries. The cheaper system made it possible for many more farmers to join milk-recording as the table shows. The accuracy of milk- recording at herd level was checked by computerized comparisons between recorded and delivered milk to the dairies instead of a person visiting the farm every month. Thus, by applying new technologies and the most efficient ways of recording, a high farmer participation rate has been achieved. This is very important for the possibilities of a high genetic improvement of the whole dairy cattle population.

The Table 1 and Figure 4 also shows that the production development has been dramatic. A doubling of the production per recorded cow has been achieved in about 40 years at the same time as many more cows became included. Practically the same development took place in the two major breeds. The extensive development of AI and application of modern methods for genetic evaluation and selection are the main reasons for the genetic improvement, which accounts for more than half of the increased production level. Already in 1970 more than 85% of the recorded cows were bred artificially, and nowadays 99 % of the recorded cows are AI bred. Thus, very effective means of disseminating improved genetic stock to all producers are in place.

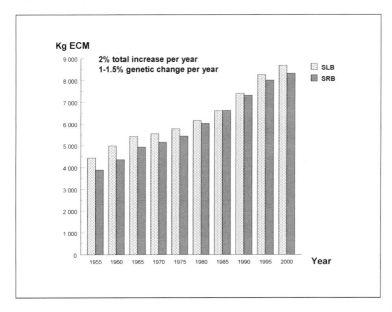

Figure 4. Annual milk production for SRB and SLB cows.

Development of breeding objectives to include functional traits into a "Total Merit Index"

By tradition selection has since long been based on production and conformation of cows. However, the more the cows produce the more important it is to directly select for traits that influence the longevity of the cows. Such traits as fertility and mastitis resistance are very important but are genetically unfavourably correlated to production. They are therefore absolutely necessary to include in the selection program if cow fertility and health is going to

Knowledge transfer in cattle husbandry

be maintained in the cow population. The problem is, however, that these traits have low heritability, although their genetic variation is considerable. Therefore rather large progeny groups, 100-150 daughters, are necessary for accurate estimation of the bull's breeding value.

Any efficient dairy cattle breeding program assumes that all farmers use young bulls to a certain extent in order to maximise the long-term genetic progress. The smaller the population, the more important it is to have a high proportion of young bulls used. This is especially important when considering functional traits, as rather large progeny groups are needed. In the Swedish situation about 30 % of the inseminations are done by young unproven bulls, a figure that is comparatively high. It reflects that farmers have understood the long-term benefits of a breeding program and, thus, voluntarily leave a significant portion of their herds available for testing purposes.

Already in 1973 a first Total Merit Index (TMI) was introduced for selection of AI bulls in Sweden. This index was developed in collaboration between scientists and progressive commercial breeders. It was indeed much less complete than the index today, but it presented a philosophy of combining all available information of the cow population to be used for selection of bulls according to the priorities of the dairy producers. Today the TMI includes the following traits:

- Production (Yields of protein and fat with a neg. weight on milk yield)
- Beef traits (Growth and carcass data from male progeny)
- Female fertility (No. ins., days calving-first AI, heat score, reprod. treatments; heifers and two first lact.)
- Calving traits (Calving difficulty and stillbirth as calf and dam traits in first-calvers)
- Udder health (Vet. treatment of clinical mastitis, somatic cell count, culling for mastitis)
- Other diseases (Vet. treatments)
- Udder conformation (International standards)
- Body conformation, feet and legs (International standards)
- Milking speed (Interview)
- Temperament (Interview)
- Longevity (Stayability after two lactations)

Figure 5. The Swedish Bull Index.

In the early days the TMI was based on production, growth rate at performance station, daughter fertility, stillbirth rate and some conformation traits. In the mid-eighties health traits were added and in the nineties the longevity record was included. Beyond that, all existing traits have gradually been developed, by e.g. better recording or genetic evaluation methods, and consist today of a number of sub-traits, which are combined into indexes. One important simplification had to be made when constructing the TMI: all genetic correlations were considered zero. Later studies have proven this simplification to be of minimal importance, although it leads to some overestimation of the expected genetic progress in different traits.

Another principally important decision was that all sub-indexes should be published, so that the farmer could see what economic weight had been put on each individual trait. That philosophy enabled any farmer to use his own weighting factors and he could see what the effects were of changing them from the weights considered appropriate for the average farm. This policy was very important for the adoption of the TMI to be extensively used.

Recording schemes and integration of data - organisational implications

The limiting factors for genetic evaluation of all desired traits and to select for a TMI as described, and the dissemination of improved genetic stock, have been the availability of data. Traditionally milk recording, AI- and herdbook activities, and genetic evaluations, have been conducted by different organisations, which has made it difficult to utilise the data effectively.

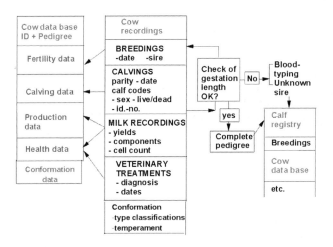

Figure 6. Integration of computerized records on pedigree, productivity and health for genetic evaluations.

Problems that frequently occur include different identification of animals, different databases, poor co-operation between organisations and poor knowledge about how data is used by other organisations. These problems were recognised 40 years ago in Sweden and thanks to the efforts by SHS, and its farmer cooperative members, it has been possible to integrate all records with unique identities of all cows into the same computer system. Milk-recording, AI-service and conformation evaluations were all done within the same farmer cooperatives, whereas health data were negotiated with the government to be included as well.

The principles of the integration of data are illustrated in Figure 6. The automated pedigree-control has completely changed the role of herd-books to be more of promotional organisations rather than responsible for certifying pedigrees.

Nowadays the database includes the following information:

- Milk recording per test-day with milk yield, fat %, protein %, cellcount and urea content
- Dates of insemination, sire identity, fertility treatments and pregnancy diagnoses
- Date of birth, calving performance, use of calf (live, stillborn, malformed), sex and identification of calf.
- Conformation traits as recorded in co-operation with AI- and herd book organisations
- Treatments of diseases by veterinarians and results from the BVD eradication program
- Date of culling, culling reasons and movement of cows. The last years recording of calving, culling and movements is done in co-operation with the government authorities responsible for the central cow database.
- Slaughter data including date of slaughter, carcass weight, carcass grades and possible disease diagnoses are recorded by the abattoirs and reported weekly to the database.

There are also some information recorded and kept mainly for breeding purposes, such as DNA-typing results for identification and for certain genotypes (mainly malformations), and specifications of malformations of individual animals.

All these data are recorded and stored on an individual cow or calf basis, and except for the conformation and DNA-typing are mainly used for management purposes. Thus, recording, processing and storing of the data are primarily financed for those reasons, and can be utilised for breeding purposes at a minimal cost.

Multi-use of the integrated data base

By integrating all data into the same computer system it is possible to develop measures on e.g. feeding, cow fertility, cow health, calving traits, calf viability and replacements to be used for breeding purposes as well as for summary statistics at herd, regional and national level, and for economic analyses, and management and extension purposes. The multi-use of the data and data-flows are illustrated in Figure 7 and 8.

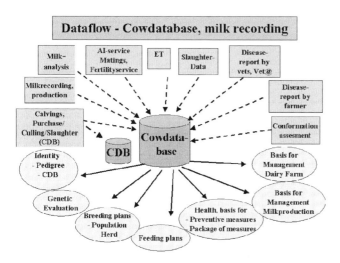

Figure 7. Dataflow – cow database, milk recording.

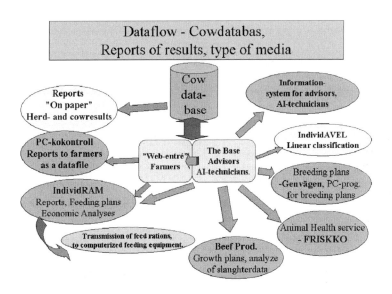

Figure 8. Dataflow – cow database, reports of results, type of media.

From a genetic point of view, it is most important to note the opportunities to use the data for research and deliver such results that can immediately be implemented and that reach all farmers participating in milk-recording and using AI. Practically all areas included in the TMI, i.e. production, fertility, growth and beef traits, calving traits, conformation and health data, have been thoroughly researched based on the field records made available, in many cases in a number of PhD-dissertations and student thesis works. Both genetic research and genetic evaluations in practice require data with accurate pedigree records for several generations. Thus, both purposes of use support each other.

An integrated database has many advantages, as all activities using cow data utilise the same information, only new information needs to be added and new records and corrections are always available for all users. This multi-use of the data means a cost effective use of information once collected. An important by-product is that the quality of the data will be improved by the many ways data are scrutinised and used. This is not least important for the research aspects of using these field data.

Genetic improvement achieved

As research results are continuously implemented in genetic evaluations at the national level, and used for selection of progeny tested bulls for wide spread use by the AI company, all farmers will be sharing the semen from the selected group of bulls for improvement of their herds. In this way the farmers gain directly from the collective efforts of their cooperative member organisations.

As previously mentioned the genetic improvement in production has been substantial. This is illustrated by the genetic trend in milk-index derived from Interbull-evaluation data for the Ayrshire group of dairy cattle including the SRB-breed (grey line —▲—).

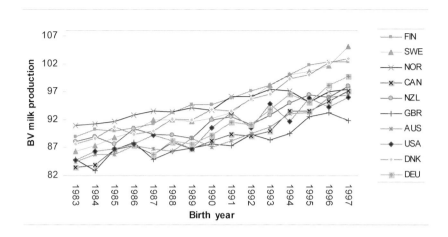

Figure 9. Genetic trends in production for SRB and Ayrshire populations from different countries (Interbull proofs on the Swedish scale).

The SRB breed shows a steadily increased genetic level for production, and the trend of nearly 2 % per year seems steeper in later years than previously (Figure 9). This is much dependent on the more efficient AI structure nowadays allowing a sharper selection of bulls, including also foreign populations. As we could expect a negative trend in cow fertility from selection for just production it is interesting to note in Figure 10 that the genetic trend of the SRB breed has been kept rather unchanged, or been slightly positive, thanks to selection of bulls according to the TMI including cow fertility. However, for Swedish Holstein breed the trend has been quite negative, because bull sires have for about two decades been selected from countries not recording or selecting for female fertility, a clear draw-back of the "holsteinisation" of our black and white cows.

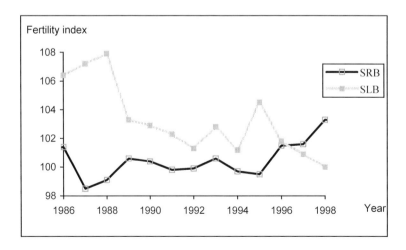

Figure 10. Genetic trends in female fertility for SRB and SLB.

Unchanged or slightly positive trends for mastitis resistance can also be shown despite the strong increases in production. Again the maintained or improved udder health is a result of the joint utilisation of veterinary records on treatments of cows for mastitis with cellcount information from the milk-recording scheme.

Farmer use of genetic information – a study of attitudes

In a recent investigation 300 Swedish dairy farmers were interviewed about their interest in breeding activities, and if and how they utilised the information and tools being available for use at herd level. A large majority, 82 %, were *very interested* or *interested* in dairy cattle breeding. Farmers with 100 cows or more were more interested than the average. Quite a high proportion of the farmers, 85 %, believe that breeding is of great importance for the economical outcome of the herd.

Two-thirds of the farmers work out mating plans for choice of individual bulls to fit the characteristics of each individual cow. The interest to get such mating plans has increased in the last few years, and about 75 % of the farmers with 100 cows or more now produce written mating plans. Today about 50 % of the mating plans are produced with a computer based optimisation programme. The programme gives suggestions of mating bulls based on the individual cow´s breeding values and the mating sires own proofs for all evaluated traits, such as calving traits, udder health, production, teat length etc.

In general the farmers show a great interest in considering health traits, specifically udder health. The farmers considered the correlation between health traits and longevity to be important. 84 % of the farmers expressed that health traits are *very important* in the breeding program, whereas conformation and production both were *very important* for 42 % of the herd owners.

More than 90 % of the farmers were reading farm/animal magazines weekly. 82 % of the farmers were on average spending 30 minutes reading the national breeding magazine after each release of new breeding values. Presently, 23 % of the farmers said they regularly visit the home pages of the breeding companies and this amount is believed to increase.

Farmer attitudes - results of an inquiry to 300 dairy producers

❏ 82% were **"Very interested"** or **"Interested"** in dairy cattle breeding

❏ 85% consider breeding (genetic) activities being of great importance for their **herd economy**

❏ Two-thirds of the farmers worked out **mating plans** for individual cows. For 50% of the mating plans are based on a computerised optimisation program

❏ 84% considered **health traits**, especially udder health, very important to consider in the breeding program - 42% for production and for conformation

❏ 82% regularly follow up the release of new breeding values of AI-bulls – 23% use the **website**

Conclusions

The power of integrated data bases for transfer of knowledge from research to the dairy farmer has been demonstrated as regards cattle breeding with examples from Swedish developments and experiences. These were based on a high participation rate among farmers and effective use of data collected in an expanded milk-recording scheme, allowing data from AI and health services to be integrated. The multi-use of an integrated data base promotes both extensive use and improved quality of the information. Thus, data are very useful for research and results can be quickly implemented into breeding programs that benefit all farmers. The organisational implications of integrated data bases are that fewer and more effective organisations can serve the farmers with improved information. The Swedish experience shows a very high involvement of the commercial dairy farmers in the breeding activities.

Producers' attitudes towards modernization and expansion

Roger W. Palmer and Jeffrey Bewley

UW-Madison Dairy Science Department, UW-Extension Dairy Team, 1675 Observatory Drive, Madison, WI 53706-1284, USA

Summary

The dairy industry throughout the world has experienced significant changes, resulting in fewer but larger dairies. New technologies allow managers to successfully operate larger dairy herds. This article uses examples of existing dairy producers who have changed from one type of production system to another. For example, producers move from "traditional dairy" systems, in which cows are housed and milked in a stall barn, to a system that may include freestall housing, TMR (total mixed rations), and milking parlor. No single system is best for everyone; therefore, producers must understand the available options and evaluate the merits of each for their operation. The profitability of a business directly influences the quality of life of its owners and workers. Profits can be used to purchase facilities, equipment, and services, which improve working conditions and support family living. Since family living expenses constantly increase, the number of animals or the profit per animal must increase to support growing family needs. Increasing product value or decreasing production costs can influence profit. Modern technologies allow producers to enhance labor efficiency, increase profits, and improve quality of life for both dairy owners and workers. Quality-of-life enhancements help preserve health and safety and often lead to better working conditions, such as more time away from the farm. These same modern technologies, however, often require large herds to decrease the investment per animal and better utilize assets. The optimal herd size varies with the operator's goals and available resources. Each producer must select and incorporate technologies that allow milk production – now and in the future – at a competitive price, and choose the management system and herd size that best provide a profitable and sustainable business. In this article Palmer and Bewley report on a questionnaire survey conducted to determine how producers and their families, who expanded their operation by at least 50 % over a five-year period, evaluated the changes they made. What was their attitude towards expansion? How satisfied are they with the different choices made in their farm-set up and how did it affect their personal and business life. Rate of satisfaction with different choices of expansion, manure removal methods, ventilation options, stocking rates of barn, bedding types, etc. are given. But also other criteria such as personal health, neighbor relations and overall quality of life are examined. Farmers were asked to share their advice based on what they had learned in the expansion process, and to define the best and worst choices they made.

Keywords: expansion, farmers' attitude, performance, management practices

Introduction

The dairy industry is going through a major restructuring. Dairy operators are incorporating modern technologies to help improve their efficiency and the quality of life of their families and workers. An extensive mail survey of dairy farms in Wisconsin, which expanded during the past five years, was conducted in the spring of 1999. The goal of this research is to determine what changes producers have made and how satisfied they are with these changes.

The results from this survey will help guide other producers considering changes to their operations.

Data provided by Wisconsin DHI was utilized to identify Wisconsin herds on DHI test that expanded between 1994 and 1998. Herds were selected if herd size had increased by at least 50 % for smaller herds (60-100 cows) or at least 40 % for larger herds (> 100 cows). To identify other herds that had expanded herd size but were not DHI members, a letter was sent to all members of Professional Dairy Producers of Wisconsin (PDPW) asking for their willingness to cooperate in the survey. Of the 694 farms that received the survey, 604 were located through DHI, 95 through the PDPW letter, and 50 through referrals (some herds were listed in multiple sources). 94 surveys were mailed on April 5, 1999 and a follow-up mailing of postcards were sent to those who had not responded by April 15. County Extension agents were asked to help producers complete the surveys or to encourage dairy producers to complete the survey.

Overall, 48 % (336) of the surveys were returned. Since some of the respondents did not meet survey criteria, returned unusable questionnaires, or declined to participate, 44 % (302) surveys were used in this analysis. Herd summary information from AgSource DHI was combined with the survey data when available. Herds were categorized based on current herd size, magnitude of expansion, and type of expansion. Herd size categories were established so that all herds were divided into five equal-sized categories. Each category contains approximately 60 herds and includes herds with 60 to 105 cows, 106-145 cows, etc. For many questions, producers were asked to indicate their satisfaction with particular aspects of their operation by choosing a number in a scale from 1 (very dissatisfied) to 5 (very satisfied). These "satisfaction values" were averaged and are summarized in the following pages. In this report, some tables will contain an "N=xx" which indicates the number of producers who responded to a question in a particular category. Means within rows or columns, that differ (P < 0.05), are denoted with superscripts a, b, c, etc.

This report provides of the results of the analysis of the survey. Hopefully it will give insight to the producers who may be considering an expansion. The results summarized indicate what facility and management options have been chosen by Wisconsin producers who have expanded and the production trends associated with these options and the producers satisfaction with their choices.

Herd Characteristics

Table 1 shows that the average herd in this survey more than doubled its size (from 102 to 252 cows) in the five-year period from 1994 to 1998. 57 % of the herds in this survey have more than 150 cows. Herd size includes milking and dry cows. By comparison, the average herd in Wisconsin had 60 cows in 1998, and the average Wisconsin DHI herd had 71 cows. Most producers in the survey appear to still be in the process of expanding as indicated by the average long-term goal of 453 cows. Figure 1 shows the distribution of these herds based on number of cows. Figure 2 demonstrates the distribution of herds based on the extent of the herd's expansion during this period.

Table 1. Average herd size of sample herds.

1994	Before Most Recent Expansion	Now	Long-Term Goal
102 Cows	136 Cows	252 Cows	453 Cows

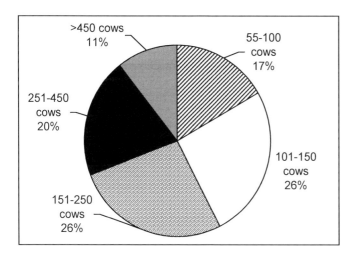

Figure 1. Herd size distribution.

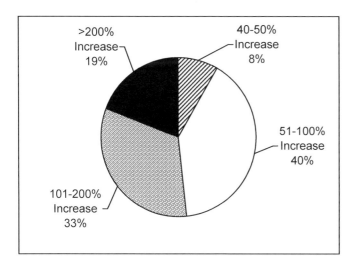

Figure 2. Expansion magnitude distribution.

Herd summary information from AgSource DHI was available for 243 of the Holstein herds in the data set. This information has been summarized in Table 2. Milk production increased for these herds from 1994 to 1998 (rolling herd average milk [RHA] increased by 1853 pounds, ME milk increased by 1957 pounds, and peak milk increased by 5 pounds). Reproductive performance seems to decrease as shown by a 0.6-month increase in calving interval and a 14-day increase in days open.

Table 2. Herd summary averages.

	1998	1994	Change
Herds, no	252	252	
Median herd size	180	80	+100
RHA milk, lbs.	21956 ± 195^a	20103 ± 187^b	+1853
ME milk, lbs.	23698 ± 198^a	21741 ± 189^b	+1957
Peak milk, lbs.	88.6 ± 0.6^a	83.6 ± 0.6^b	+5
Linear SCS	2.91 ± 0.03	*	*
Days dry	61 ± 1^b	63 ± 1^a	-2
Calving interval, months	13.8 ± 0.1^a	13.2 ± 0.0^b	+0.6
Days open	140 ± 2^a	126 ± 2^b	+14
Age at first calving, months	26.1 ± 0.1	*	*
Culling rate, %	33.2 ± 0.7	*	*

$^{a, b}$ Means within rows with different superscripts differ (P < 0.05); lbs / 2.2 = kilograms

A common question asked by producers considering an expansion is "Should I start with all new facilities or modify what I have?" Most producers (72 %) who responded to this survey indicated they used existing facilities along with some new facilities. It appears that producers using all new facilities have higher production and greater satisfaction with net farm income, personal health, disposable household income, and time away from the farm (Table 3).

Table 3. Operation performance and satisfaction by "Type of Expansion".

	No new facilities	Some new facilities	All new facilities
Herds, no.	31	218	53
1998 mean herd size	109	216	483
1994 mean herd size	61	97	145
1998 RHA milk, lbs	20503^c	21920^b	23218^a
1994 RHA milk, lbs	17985^b	20300^a	20897^a
Change in RHA	2519^a	1658^b	2321^{ab}
Linear SCS	3.07	2.89	2.82
Calving interval, months	14.0	13.6	14.0
Age at first calving, months	26.9^a	26.0^b	25.7^b
Culling rate	34.7^{ab}	33.7^a	29.2^b
Cows per FTE	30^c	38^b	52^a
Heat detection *	3.32^b	3.56^b	3.87^a
Production costs per cwt. *	3.61^b	3.66^b	4.04^a
Net farm income *	3.26^b	3.59^b	4.04^a
Neighbor relations *	4.29^a	3.95^{ab}	3.77^b
Personal health *	3.32^c	3.78^b	4.23^a
Disposable household income *	3.23^b	3.59^b	4.08^a
Time away from the farm *	2.71^c	3.29^b	3.79^a
Overall quality of life *	3.65^b	3.82^{ab}	4.06^a

$^{a, b, c}$ Means within rows with different superscripts differ (P < 0.05); lbs / 2.2 = kilograms
*Average satisfaction reported on a scale from 1 (very dissatisfied) to 5 (very satisfied).

Knowledge transfer in cattle husbandry

Producers who built all new facilities achieved the lowest average culling rate and had the highest overall farm labor efficiency based on their average cows per full-time equivalent [FTE]number of employees. Producers who did not change facility type felt they had better neighbor relations than those who built all new facilities. It is interesting to note that producers expanding without adding new facilities saw the largest production increase.

Another frequently asked question among producers considering expansion is "How big should I get?" As mentioned earlier, herds were divided into five categories by herd size (Table 4). The average 1998 herd size for the herds in each of these categories were 86, 126, 183, 272 and 597, respectively. The average milk production per cow increased as herd size increased. The average rolling herd average for the larger herds was 4347 pounds higher than the smaller herds.

Table 4. Operation performance and satisfaction by "Herd Size".

Herd Size (number of cows)	60-105	106-145	146-220	221-360	> 360
Herds, no.	61	62	59	60	60
1998 mean herd size	86	126	183	272	597
1994 mean herd size	47	71	90	107	197
1998 RHA milk, lbs	19766^d	21642^c	22370^{bc}	22737^b	24113^a
1994 RHA milk, lbs	18136^d	19643^c	20690^b	20894^{ab}	21998^a
Change in RHA	1660^a	2017^a	1725^a	1843^a	2115^a
Linear SCS	3.03^a	2.96^{ab}	2.83^b	2.85^{ab}	2.80^b
Days open	130^b	136^{ab}	143^a	136^{ab}	143^a
Days dry	64^a	61^{ab}	60^b	61^{ab}	61^{ab}
Age at first calving	26.8^a	26.2^a	26.3^a	25.4^b	25.2^b
Culling rate	31.5	33.5	33.4	35.1	32.4
Cows per FTE	27^c	34^b	40^b	49^a	51^a
Acres per cow	3.38^a	3.37^a	2.64^b	2.61^b	2.31^b
Milk production level *	3.53^b	3.55^b	3.97^a	3.82^{ab}	3.92^a
Heat detection *	3.48^b	3.37^b	3.60^{ab}	3.58^{ab}	3.92^a
Conception rate *	3.39^{ab}	3.23^b	3.48^{ab}	3.22^b	3.55^a
Calving interval *	3.44^{ab}	3.26^b	3.50^{ab}	3.37^{ab}	3.62^a
Milk quality *	3.64^{ab}	3.58^b	3.91^a	3.73^{ab}	3.88^{ab}
Production costs *	3.53^{bc}	3.50^c	3.83^{ab}	3.71^{bc}	4.03^a
Net farm income *	3.26^b	3.33^b	3.77^a	3.73^a	4.10^a
Neighbor relations *	4.13^a	3.97^{ab}	3.88^{ab}	4.02^{ab}	3.78^b
Personal satisfaction with my role *	3.98^b	3.92^b	4.03^{ab}	4.10^{ab}	4.30^a
Personal health *	3.38^c	3.65^{bc}	3.71^{bc}	4.02^{ab}	4.30^a
Disposable household income *	3.33^c	3.39^c	3.55^{bc}	3.87^{ab}	4.12^a
Relationship with spouse and family *	4.08^{ab}	3.80^b	4.07^{ab}	4.17^a	4.08^{ab}
Time away from the farm *	2.85^b	3.02^b	3.22^b	3.63^a	3.88^a
Overall quality of life *	3.75^{bc}	3.60^c	3.69^c	4.03^{ab}	4.13^a

$^{a, b, c, d}$ Means within rows with different superscripts differ (P < 0.05);   lbs / 2.2 = kilograms
*Average satisfaction reported on a scale from 1 (very dissatisfied) to 5 (very satisfied).

Production increases from 1994 to 1998 was similar across all size categories (Table 4). Larger herds tended to calve animals at a younger age than the smaller herds. The overall farm labor efficiency, as measured in cows per full time equivalent, increased with herd size. The largest and smallest herds averaged 51 and 27 cows/FTE, respectively. Part of this difference is due to the larger number of acres of land farmed by the smaller herds, 3.38 and 2.31 acres per cow. Larger producers appear to be more satisfied than their small herd counterparts in all areas except neighbor relations. Larger producers reported significantly higher satisfaction with net farm income, personal satisfaction with their role, personal health, disposable household income and time away from the farm. It is interesting that when they were asked about their overall quality of life the largest and smallest herd size groups reported the highest ratings. The herds in the intermediate size (106-145 and 146-220 cows) tended to be less satisfied. When operation performance and satisfaction were summarized by the magnitude of expansion, results were very similar to the herd size summary (i.e. herds with the greatest magnitude of expansion had high production and appeared to be more satisfied with income and personal satisfaction).

Facilities

Seventy two percent of the producers in this survey used a combination of modified and new facilities, while 18 % built all new facilities with the remaining producers making no major changes of facility type. Not surprisingly, the primary change in housing type was from the traditional stanchion or tie-stall barn to modern freestall barns. Figure 3 shows the distribution of herds by barn type and by freestall barn type.

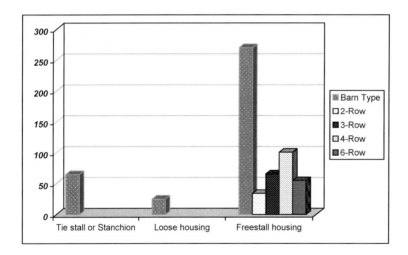

Figure 3. Housing for the milking herd.

Table 5 shows that for all herds with freestall barns, the 4-row design had the highest average milk production average per cow (22,938 pounds), lowest average linear score (2.78), and highest feed intake satisfaction (4.40), along with having the highest stocking rate (111%). These numberic averages, although not statisitically significant, show a trend that becomes more obvious when only new freestall barns with drive-through feeding are examined. Cost per stall varies by barn type and is influenced by combining new and remodeled facilities.

Knowledge transfer in cattle husbandry

Table 5. Freestall barn type.

	2-Row	3-Row	4-Row	6-Row
Herds, no.	17	60	96	55
1998 median herd size	144	130	240	320
1998 RHA milk, lbs	22291[ab]	21528[b]	22938[a]	22469[ab]
1994 RHA milk, lbs	19861[ab]	19765[b]	20994[a]	20852[ab]
Change in RHA	2431	1764	1944	1617
Stocking rate (%)	104	103	111	104
Linear SCS	2.82	2.87	2.78	2.97
Cows per FTE	35[b]	38[b]	44[ab]	47[a]
Cow comfort *	4.00[b]	4.67[a]	4.54[a]	4.50[a]
Cow cleanliness *	3.88[b]	4.50[a]	4.33[a]	4.39[a]
Feed intake *	4.29	4.21	4.40	4.35
Ability to move animals *	4.12[ab]	4.11[b]	4.30[ab]	4.57[a]
Feeding convenience *	3.82[b]	4.53[a]	4.55[a]	4.74[a]
Cost per stall	$712[b]	$908[b]	$1120[a]	$1215[a]

[a, b] Means within rows with different superscripts differ (P < 0.05);   lbs / 2.2 = kilograms
*Average satisfaction reported on a scale from 1 (very dissatisfied) to 5 (very satisfied).

Herd performance and producer satisfactions seem to differ between freestall feed delivery designs (Table 6). Producers who chose barns with drive-through feeding were more satisfied with feeding convenience and manure management than those that chose other options. Their 1998 milk production was significantly higher compared to outside feeding. Their overall farm labor efficiency (45 vs. 37 cows/FTE) and ability to move animals was better than producers with drive-by facilities. The average cost per stall was highest for the drive-through group, which in part may reflect the fact that many of the drive-by and outside feeding barns were retro-fits of existing structures.

Table 6. Freestall feed delivery design.

	Drive-through	Drive-by	Outside Feeding
Herds, no.	155	44	19
1998 median herd size	245	144	125
1998 RHA milk, lbs	22657[a]	21608[ab]	20981[b]
1994 RHA milk, lbs	20941[a]	19711[b]	18719[b]
Change in RHA	1715	1897	2262
Cows per FTE	45[a]	37[b]	40[ab]
Feed intake *	4.38	4.14	4.26
Ability to move animals *	4.38[a]	4.02[b]	4.16[ab]
Manure management *	3.94[a]	3.55[b]	3.26[b]
Feeding convenience *	4.73[a]	4.36[b]	3.74[c]
Cost per stall	$1196	$877[b]	$522[c]

[a, b, c] Means within rows with different superscripts differ (P < 0.05);   lbs / 2.2 = kilograms

Like Table 5, Table 7 summarizes production and satisfaction values for different freestall barn types except only newly constructed barns with drive-through feeding were included. By adding these qualifications, bias created by including remodeled facilities or other feeding strategies is removed. Similar trends are shown as in Table 5 except for cost per stall of different barn types. 1998 Milk production per cow was significantly higher for the 4-row barn (+1911 pounds per cow per year) and the average linear somatic cell count score was lower. The 4-row barns were stocked heavier (9 % more) and producers reported higher satisfaction with feed intake and cow comfort. Interestingly, producers reported spending only a few dollars more per stall (+ $23/stall) for their 4-row barns than those that chose the 6-row option. Since these barns had a higher stocked rate, the actual cost per cow was less ($1103 vs. $1177 per cow).

Table 7. Freestall barn type (New barns with drive-through feeding only).

	4-Row	6-Row
Herds, no.	53	42
1998 median herd size	245	247
Rolling Herd Average (RHA) 1998	23,644[a]	21,733[b]
Change in RHA	1974	1382
Stocking rate (%)	112[a]	103[b]
Average Linear Score	2.73[b]	2.96[a]
Feed intake satisfaction *	4.47	4.33
Cow Comfort *	4.66	4.45
Cost per Stall	$1235	$1212

[a, b] Means within rows with different superscripts differ ($P < 0.05$).

Selection of freestall bases, surfaces, and bedding types is very important. Often producers need to choose between sand and mattress based freestalls. The DHI records, for the producers who responded to this survey, show no significant difference in milk production or somatic cell counts between those using sand or mattresses after their expansion (Table 8).

Table 8. Freestall bedding type.

	Mattresses	Sand
Herds, no.	69	145
1998 median herd size	265	195
1998 RHA milk, lbs	22519	22539
Change in RHA	1587	2071
Linear SCS	2.88	2.80
Cows per FTE	45	40
Culling rate (%)	34	32
Cow comfort *	4.42	4.55
Cow cleanliness *	4.12[b]	4.47[a]
Hock damage *	4.22[b]	4.72[a]
Teat damage *	4.48	4.59
Udder health *	4.09	4.31
Bedding usage and cost *	4.25[a]	3.95[b]
Manure management *	4.32[a]	3.43[b]
Cost per stall	$1306[a]	$946[b]

[a, b] Means within rows with different superscripts differ ($P < 0.05$);   lbs / 2.2 = kilograms
*Average satisfaction reported on a scale from 1 (very dissatisfied) to 5 (very satisfied).

Producers using sand seem to be more satisfied with cow comfort issues, and less satisfied with manure management and bedding issues than those using mattresses. Sand users reported significantly higher satisfaction scores for cow cleanliness and hock damage, whereas mattress users reported significantly higher satisfaction with bedding use and cost and manure management. The average cost per stall was higher for mattress based stall users, part of which would be the initial cost of the mattress. Although not significantly different, herds with mattress based stalls seem to have higher overall labor efficiency (45 vs. 40 cows/FTE). Culling rates were similar with a slight numeric advantage to sand users (Table 8).

Table 9 examines the effect of overstocking of all new freestall barns. Herds were categorized by stocking rate. Interestingly there was no significant difference in 1998 RHA milk production and a trend toward higher production with higher stocking rates. Part of this effect may be explained by results shown in Table 7, which indicated that herds with 4-row barns were being stocked at a higher rate and had higher milk production levels. When 4-row barn users were analyzed separately, their production did not change significantly with stocking rate. An analysis of culling rate did not show any significant difference in culling level (31, 38, 32, 31, 31 and 30 % respectively) for the categories shown. The cost per stall reported did not differ by category, but when cost per cow was calculated using stocking rate information it varied substantially.

Table 9. Stocking rate of new freestall barns.

	Under capacity	At capacity	1-10 % over crowded	11-20 % over crowded	21-30 % over crowded	>30 % over crowded
Herds, no.	46	4	46	37	16	8
1998 median herd size	128	115	263	275	373	230
1998 RHA milk, lbs.	21501	20693	22743	22768	22244	23297
1994 RHA milk, lbs.	19621[b]	19123[ab]	21075[a]	21128[a]	20032[ab]	21682[ab]
Change in RHA	1879	1570	1668	1640	2211	1615
Cow comfort *	4.49	5.00	4.57	4.65	4.44	4.50
Feed intake *	4.33	5.00	4.35	4.43	4.50	4.13
Cost per stall	$1076	$1183	$1162	$1174	$916	$1192
Cost per cow	$1266[a]	$1183[abc]	$1094[ab]	$988[bc]	$708[c]	$793[abc]

a, b, c Means within rows with different superscripts differ (P < 0.05);   lbs / 2.2 = kilograms
*Average satisfaction reported on a scale from 1 (very dissatisfied) to 5 (very satisfied).

Table 10. Use of fans and sprinklers.

	Fans and Sprinklers	Fans	Sprinklers	Neither
Herds, no.	28	59	18	139
1998 median herd size	243	283	203	190
1998 RHA milk, lbs	22964[a]	23943[a]	23381[a]	21800[b]
1994 RHA milk, lbs	20841	21309	20993	20183
Change in RHA milk	2123	2634	2388	1627
Linear SCS	2.82	2.71	2.78	2.87
Cow comfort *	4.42	4.33	4.29	4.56
Cow cleanliness *	4.14[b]	4.39[ab]	4.32[ab]	4.43[a]

a, b Means within rows with different superscripts differ (P < 0.05);   lbs / 2.2 = kilograms
*Average satisfaction reported on a scale from 1 (very dissatisfied) to 5 (very satisfied).

Table 10 examines the use of fans and sprinklers in freestall barns to cool animals. Only 37% of producers report using fans and only 18 % report using sprinklers. Average 1998 RHA milk was significantly higher for producers who used either or both fans and sprinklers (+1164, +2143 and +1581 pounds, respectively for the categories shown). Interestingly producers using cooling equipment didn't report higher satisfaction with cow comfort and producers not using cooling equipment were more satisfied with cow cleanliness that those using both.

Table 11 compares manure removal methods. Almost 80 % of respondents reported using a tractor to scrape barns as their primary manure handling method. Their reported freestall barn cost per stall was significantly less than producers selecting slats and numerically less than alley scraper or flush removal methods. Overall farm labor efficiency, as measured in cows per FTE, was significantly higher for alley scrapers.

Table 11. Manure removal method.

	Tractor Scrape	Alley Scrapers	Slats	Flush
Herds, no.	189	26	17	5
1998 median herd size	205	283	370	545
Cows per FTE	42^b	50^a	43^{ab}	38^{ab}
Manure management *	3.55^b	4.39^a	4.65^a	5.00^a
Bedding usage and cost *	3.95^b	4.39^a	4.41^a	4.20^{ab}
Cost per stall	986^b	1111^b	1458^a	1095^{ab}

$^{a, b}$ Means within rows with different superscripts differ (P < 0.05).
*Average satisfaction reported on a scale from 1 (very dissatisfied) to 5 (very satisfied).

As herds expand, feed storage requirements increase. Table 12 shows the satisfaction with different feed storage types. Satisfaction with bunkers or trenches appears to be much higher (4.37) than with upright silos (3.54) or flat pads (3.43) and storage bags (3.73).

Table 12. Feed storage.

Feed Storage System	Upright silos	Flat pads for piles	Bunkers or trenches	Storage bags
Satisfaction *	3.54^b (N=253)	3.43^b (N=108)	4.37^a (N=156)	3.73^b (N=171)

$^{a, b}$ Means within rows with different superscripts differ (P < 0.05).
*Average satisfaction reported on a scale from 1 (very dissatisfied) to 5 (very satisfied).

Table 13 Feed acquisition strategies and acres per cow farmed.

		Forages			
		Raised all	Raised most	Bought Half	Bought Most
Grains	Raised all	3.99 (N=90)	3.28 (N=17)		
	Raised most	2.79 (N=28)	2.80 (N=34)	2.21 (N=1)	
	Bought half	2.32 (N=19)	2.10 (N=23)	2.14 (N=3)	
	Bought most	2.08 (N=7)	1.97 (N=19)	1.89 (N=5)	
	Bought all	2.40 (N=14)	1.94 (N=13)	1.09 (N=10)	1.05 (N=4)

Table 13 shows the reported feed acquisition strategy and a calculated value of acres per cow of land farmed by respondents. Producers who reported raising all of their forage and grain farmed, on the average, 3.99 acres per cow. At the other extreme were producers who reported buying most of their forage, all of their grain and averaged 1.05 acres per cow.

Most of the Wisconsin producers that expanded or are thinking about expanding their dairy operation had or have a stall barn with pipeline milking system and must change their milking system to support the larger herd size. Table 14 lists the milking system options normally considered by these dairy producers and are listed by their relative capital investment cost. Most (59 %) of the producers in this study reported using some type of a pit parlor while 22% milk in a traditional stall barn. Of those with pit parlors, 58 % milk in parallel parlors and 37% have herringbone parlors. Flat barn parlors were predominately of the walk-through type (79 %). Herds selecting the three cheaper options were generally smaller in size and probably chose those options to keep capital investment per cow down. Pit parlors in a new building were chosen by herds that were larger before expansion and were probably in their second phase of the expansion process and could justify the additional costs associated with building a new parlor complex.

Table 14. Milking facility performance.

	Stall barn with pipeline	Flat parlor in old barn	Pit parlor in old barn	Pit parlor in new building
Herds, no.	65	52	73	107
1998 mean herd size	117	157	212	411
1994 mean herd size	62	71	95	148
1998 RHA milk, lbs	20684^c	21397^{bc}	22207^{ab}	23073^a
Change in RHA	1929	1773	1721	2019
Linear SCS	3.02^a	2.97^{ab}	2.86^{ab}	2.78^b
Cows per FTE	29^c	38^b	43^{ab}	45^a
Number milking units	7^d	9^c	14^b	20^a
Number workers	2.4^a	2.2^{ab}	2.0^b	2.1^{ab}
Time to milk one shift, hr	2.21^c	2.64^{bc}	3.04^b	4.15^a
Cows per hour	47^c	55^{bc}	61^b	83^a
Cows per worker hour	21^d	27^c	34^b	43^a
Time spent milking *	3.03^c	3.92^{ab}	3.78^b	4.12^a
Physical comfort of milker*	2.45^c	3.83^b	4.10^{ab}	4.32^a
Milk quality *	3.28^b	3.75^a	3.66^a	3.70^a
Cleanliness and ease of setup *	3.55^b	3.54^b	3.75^{ab}	3.97^a
Safety of operator *	3.31^c	3.40^c	4.01^b	4.38^a
Cost per milking unit	\$4191^b	\$4954^b	\$6500^b	\$15832^a

$^{a, b, c, d}$ Means within rows with different superscripts differ (P < 0.05);   lbs / 2.2 = kilograms
*Average satisfaction reported on a scale from 1 (very dissatisfied) to 5 (very satisfied).

Labor efficiency in this study was measured in two ways. The overall efficiency of the farm operation was expressed as cows per full time equivalent and the milking system's labor efficiency by cows per worker hour. Labor efficiency increased with each category and was the highest for the pit parlor in new building option. The average cows per worker hour was 21 for stall barn with pipeline systems, 27 for flat parlors in old barn, 34 for pit parlor in old barn and 43 for pit parlor in new building. This increase in parlor efficiency appeared to relate to the overall efficiency of the farming operation since the cows per FTE increased with each milking system type and only the difference between the pit parlor in old barn and pit parlor in new building were not significantly different. Very little difference was found in the number of workers used for milking. All milking system types averaged between 2.0 and 2.4 workers and only the pit parlor in old barn was significantly different than stall barn with pipeline users.

Table 14 also indicates the somatic cell count level of herds with pit parlors in new buildings were significantly lower than milking in a stall barn with pipeline, but not significantly lower than flat barn parlors or pit parlors in old barns. The number of milking units increased with each parlor type. Pit parlors were the largest with 20 units per farm. On the average, cows per milker unit tend to increase with each milking facility type (16.7, 17.4, 15.1 and 20.6, respectively). This ability to more fully utilize the milking facility is a characteristic of highly profitable farms. On the average, time spent milking for these different milking facility types (3.03 to 4.12 hours per day) indicates most producers have overbuilt their milking facility for current needs and have excess milking capacity to support future herd size growth. Producers milking in pit parlors were more satisfied with time spent milking, milk quality and safety of operator than people milk in stall barn with pipeline. Flat barn parlors were preferred over stall barn with pipeline for time spent milking, physical comfort of the milker and milk quality.

Table 15. Pit parlors.

	Auto-Tandem	Herringbone	Parallel
Herds, no.	7	67	104
1998 median herd size	190	275	245
1998 RHA milk, lbs	22146	22715	22721
1994 RHA milk, lbs	20143	20984	20840
Change in RHA	2003	1832	1880
Linear somatic cell score	2.76	2.81	2.82
Cows per FTE	33	43	46
Number stalls	10[b]	18[a]	19[a]
Number milking units	10[b]	17[a]	18[a]
Number workers	2.1	2.0	2.2
Time to milk one shift, hr	2.93[ab]	4.16[a]	3.48[b]
Extra time to set-up and clean-up, hr	0.63	0.82	0.81
Cows per hour	60	71	75
Cows per worker hour	30	38	40
Turns per hour	6.20[a]	4.13[b]	4.44[b]
Time spent milking *	4.00	3.89	4.02
Physical comfort of milker*	4.71	4.05	4.30
Milk quality *	3.71	3.74	3.62
Cleanliness and ease of setup *	4.14	3.85	3.88
Safety of operator *	4.14	3.85	3.88
Cost per milking unit	$17268[a]	$8944[b]	$13201[a]

[a, b] Means within rows with different superscripts differ (P < 0.05);　　lbs / 2.2 = kilograms
*Average satisfaction reported on a scale from 1 (very dissatisfied) to 5 (very satisfied).

Most producers using pit parlors reported using herringbone or parallel parlor designs (Table 15). A few producers reported using the auto-tandem parlor design. Milk production levels, somatic cell count levels, labor efficiency levels, and user satisfaction scored were not significantly different. This is partially due to the small number of auto-tandem parlors and the similarity in performance expected from parallel and herringbone parlors. Auto-tandem parlors had less milking units, higher turns per hour and cost more per milking unit which would be expected of a milking system that handles animals individually rather than as groups. The average cost per stall of herringbone parlors was less than parallel parlors and was probably caused by a higher percentage of herringbone parlors in old buildings and the availability of used herringbone equipment. Although not significantly significant, parallel parlors tended to have high labor efficiency values.

Animal Management Practices

A primary objective of this survey was to determine user satisfaction with different animal restraint systems (Table 16). Both self-locking manger stalls and palpation rails received high ratings for ease of use, comfort of worker, and labor efficiency. Only a few herds (27) reported only using palpation rails, whereas, 102 herds reported using only self locks to handle their animals. The only satisfaction difference was relating to initial cost.

Table 16. Animal restraint systems factors.

	Self-locking manger stalls	Palpation rails
Herds, no.	102	27
1998 median herd size	349	276
1998 RHA milk, lbs.	22,962	23,244
Change in RHA	1787	1954
Cows per FTE	45	44
Feed intake *	4.38	4.31
Initial Cost *	3.71^b	4.73^a
Ease of use/comfort of worker *	4.52	4.42
Labor efficiency *	4.51	4.38
Worker safety *	4.40	4.54

[a, b, c, d] Means within rows with different superscripts differ (P < 0.05);   lbs / 2.2 = kilograms
*Average satisfaction reported on a scale from 1 (very dissatisfied) to 5 (very satisfied).

Table 17. Maternity area satisfaction.

	Individual calving pens	Group calving on bedding pack
Herds, no.	80	105
1998 median herd size	200	245
1998 RHA milk, lbs.	22784	22701
Change in RHA milk	2039	1938
Cows per FTE	40^b	46^a
Cow health *	3.93	3.69
Calf health *	4.04^a	3.60^b
Labor use *	3.83	3.63

[a, b] Means within rows with different superscripts differ (P < 0.05);   lbs / 2.2 = kilograms
*Average satisfaction reported on a scale from 1 (very dissatisfied) to 5 (very satisfied).

Table 17 shows the production and user satisfaction with calving cows in individual pens or on a group calving bedding pack. The only significant producer perception difference was an advantage for calf health using individual pens. The overall labor efficiency dairy farms, as defined by cows per FTE, was higher for herds using group calving (40 vs. 46 cows/FTE).

Table 18 shows the percentage of herds using individual pens or group calving on bedding pack to freshen animals. Herds with 106 to 145 cows seem to be most likely to use individual calving pens and the use of a group calving area to freshen cows seems to increase with herd size.

Table 18. Maternity area by herd size*.

Maternity area type	60 to 105	106 to 145	146 to 220	221 to 360	>360
Individual calving pens	33 %	48 %	33 %	34 %	35 %
Group calving on bedding pack	19 %	33 %	50 %	53 %	70 %

* Includes herds which selected having "most" or "all" animals calve in a particular location.

Table 19 demonstrates that as herds become larger they appear to utilize custom heifer raisers more to rear their replacements. Twenty nine percent of herds larger than 360 cows have most or all of their heifers custom raised while only 7 % of herds with 60 to 105 cows use custom heifer raising to this degree.

Table 19. Custom heifer raising by herd size.

Amount	60 to 105	106 to 145	146 to 220	221 to 360	>360
Most or All	7 %	22 %	28 %	28 %	29 %

As herds expand, they must determine how additional animals will be obtained. Table 20 shows that most expanding herds purchased heifers before they calved (66 %) or bought mature animals (63 %). The practice of buying heifers "after they have calved" is interesting to note since 21 % of producers reported using this practice.

Table 20. Type of animals purchased.

Where did the additional animals come from?	No. Herds	%
Bought bred heifers before they calved	198	66 %
Bought mature animals	188	63 %
Grew from within	145	48 %
Bought bred heifers which had recently calved	64	21 %
Bought calves/heifers and raised them	50	17 %

Bio-security measures taken to prevent disease introduction by purchased animals is shown in Table 21. It appears that most producers are visually inspecting the animals (91 %), vaccinated their existing herd (67 %) and vaccinated incoming cattle (51 % after and 49 % before movement) while fewer producers are isolating (27 %) and blood testing (21 %) incoming animals.

Table 21. Biosecurity practices used.

What practices were used to minimize health problems with new animals?	No. Herds	%
Visually inspected animals before purchase	238	91
Increased level of vaccination in existing herd	177	67
Vaccinated incoming cattle after moving them	134	51
Vaccinated incoming cattle before moving them	129	49
Examined individual somatic cell count records	110	42
Isolated animals after moving them	72	27
Examined individual cow health records	67	26
Blood tested animals before purchase	56	21
Did bulk tank cultures before purchase	39	15

Table 22 shows the average 1998 DHI RHA milk production for herds milking two times or three times per day and herds reporting using or not using bST in their herd. Sixty eight percent of the herds in the survey reported using bST on an average of 58 % of the milking herd. Average milk production is highest for herds milked 3x and use bST and lowest for herds milked 2x and not using bST. The differences shown imply a 12.8 to 14.3 % increase from 3x milking and a 9.8 to 11.3 % increase from the use of bST, with an additive effect of about 24.1 % for both practices.

Table 22. Milking frequency and bST use.

RHA Milk	2x	3x	2x-3x Difference
No bST	19,830[c]	22,672[ab]	+2842 (+14.3 %)
bST	22,078[b]	24,607[a]	+2529 (+12.8 %)
bST Difference	+2248 (+11.3 %)	+1935 (+9.8 %)	+4777 (+24.1 %)

[a, b, c] Means within rows with different superscripts differ ($P < 0.05$).

Table 23. Herd performance by level of AI use

	Level of AI Use			
	All natural	Mostly natural	Mostly AI	All AI
Herds, no.	14	12	71	143
1998 median herd size	222	186	349	206
1994 median herd size	67	80	90	76
1998 RHA milk, lbs	21319[ab]	19677[b]	22069[a]	22075[a]
1994 RHA milk, lbs	18329[b]	17931[b]	20462[a]	20245[a]
Change in RHA	3088[a]	1746[ab]	1607[b]	1868[b]
Calving interval, months	13.0[b]	13.7[ab]	13.9[a]	13.7[a]
Days open	122[b]	129[ab]	144[a]	136[ab]
Days dry	59[ab]	65[a]	64[a]	60[b]

[a, b] Means within rows with different superscripts differ ($P < 0.05$); lbs / 2.2 = kilograms

In this survey, 42 % of producers indicate using a bull in their breeding program. Table 23 shows production and reproductive performance of herds based on their level of AI use. 1998

DHI milk production was significantly higher for herds using all or mostly AI than herds using mostly natural service. Herds using all or mostly natural service had significantly larger changes in milk production averages between 1994 and 1998. Calving interval is significantly lower for herds using all natural service than herds using mostly or all AI. Care should be exercised when reviewing these data since such a small number of the herds use a high percentage of natural service compared to those using a high percentage AI. As a generalization the numbers do tend to indicate higher production levels for herds using mostly AI and some reproductive advantages to herds using natural service.

Labor

This survey shows an increasing number of full-time (1.29 to 6.89) and part-time (1.48 to 4.85) hired employees with increasing herd size (Table 24). However, there no significant difference between herds concerning hours worked per week per person (46 to 52). Farms with larger herds appear to be achieving better labor efficiency since yearly hours per cow decreases from 111 for smaller herds to 56 for larger herds and cows per full time equivalent increased from 27 to 51 cows. Herds with 221 to 360 and greater than 360 cows had significantly higher cows per FTE and lower yearly hours per cow than other size groups. Part of the difference in the overall labor efficiency of farms may be caused by the amount of cropping done by each group. The average acres per cow for each group shows significantly less acres per cow for larger herds. Although this may be a factor in the difference in overall labor efficiency it is doubtful that it explains the large differences found.

An analysis of producer satisfaction questions showed that owners of larger herds felt they and their family spent less time doing farm work and more time hiring, training and managing employees. Larger herds were more satisfied with their ability to find good farm employees, training and supervising them. They were also happier with their ability to get the necessary farm work done.

Table 24. Labor usage.

Labor Related Factors	60 to 105	106 to 145	146 to 220	221 to 360	> 360
Number Family Members	2.93 (N=61)	3.10 (N=62)	3.20 (N=59)	3.28 (N=60)	3.52 (N=58)
Number Full-time Employees	1.29 (N=7)	1.38 (N=29)	2.03 (N=36)	2.62 (N=47)	6.89 (N=55)
Number Part-time Employees	1.48 (N=23)	1.84 (N=38)	2.91 (N=40)	3.18 (N=50)	4.85 (N=47)
Total Hours per Person per week	52	48	46	46	48
Yearly Hours per Cow	111^a	84^b	72^c	60^d	56^d
Cows per Full Time Equivalent	27^c	34^b	40^b	49^a	51^a
Acres per cow	3.3^a	3.3^a	2.6^b	2.6^b	2.3^b

[a, b, c, d] Means within rows with different superscripts differ (P < 0.05).

Table 25 shows the average wages reported by survey respondents for different employee classifications. Full time herd managers were the highest paid employees. For employees paid on a monthly basis, full-time non-milkers were the next highest paid followed by full-time milkers, other part-time workers, and part-time milkers. For employees paid on a hourly basis,

Knowledge transfer in cattle husbandry

full-time milkers were the next highest paid followed by full-time non-milkers, part-time milkers, and other part-time employees. Managers tended to be paid monthly salaries rather than hourly like other job classifications. Established employees received higher wages than new employees.

Table 25. Average wages.

	New Employees				Established Employees			
	$/hr	N	$/mo	N	$/hr	N	$/mo	N
Managers (full time)	$8.48	22	$2275	34	$10.58	21	$2307	54
Non-milkers (full time)	$7.18	76	$1762	21	$8.53	77	$2019	27
Milkers (full time)	$7.32	131	$1596	16	$8.87	132	$1779	29
Milkers (part time)	$6.80	162	$750	4	$7.96	147	$984	8
Other (part time)	$6.37	113	$963	4	$7.59	104	$950	4

Table 26 shows the types and frequencies of benefits provided to employees of surveyed herds. The most common benefits are paid vacations and health insurance.

Table 26. Employee benefits.

Benefits provided to full-time employees	Number Herds
Paid vacation time	144
Health Insurance	143
Milk or Meat	107
Housing	89
Other	38
Profit-sharing	24
Allow employee owned animals in herd	20
Retirement Plan	19
Share of calves born	7

Expansion

Table 27 shows the response frequency of survey respondents when asked their most important reason for expansion.

Table 27. Reasons for expansion.

Why did you decide to expand your herd? *	No. Herds	%
To increase our farm's profitability	265	89 %
To improve labor efficiency	217	73 %
To improve physical working conditions for operators	207	69 %
To get time away from the farm (by allowing us to bring in more hired help)	181	61 %
To allow a family member to join the operation	103	34 %
Other	52	17 %

*Percentages do not add up to 100 % because multiple answers could be selected.

The most frequent response was to increase their farms profitability. The other responses are listed below. It is interesting that 34 % of the herds expanded to allow a family member to join the operation. This is a large percentage considering the fact that many families do not have family members at a career decision age (Table 27).

Table 28 shows the response to the question posed to respondents, if another producer asked you about your expansion project, "Knowing what you know now, would you do it again, how would you respond?" This question was overwhelmingly answered in the positive. All but 6 % of the respondents would do it again. Of them, 66 % would do it the same way and over half would do it quicker and/or bigger.

Table 28. Satisfaction with expansion choice.

If another producer asked you about your expansion project, "knowing what you know now, would you do it again?". How would you respond? Answer given:	Number of herds	% of herds
Yes, the same way	148	51 %
Yes, only quicker	84	29 %
Yes, only bigger	66	23 %
Yes, but slower	17	6 %
No	16	6 %

Producer Comments

The survey was designed to collect categorical and numerical data that could be analyzed. However, the "comments" section of the survey has proven to be very interesting. This comments area allowed the user to write anything felt to be of importance. Comments were manually arranged by categories. When respondents wrote about more than one topic, their comments were placed in multiple categories. Below, you can see the top responses to those questions with the number of respondents for each category listed in parenthesis. Almost all of the advice given has to do with the planning process. Defining family goals, using of outside consultants and keeping an open mind are underlying messages from those that have been through the process.

Table 29. Advice on expansion.

Based on what you learned what advice would you give others considering expansion?	
Plan, Plan, Plan. Consider future needs. Research, do homework (N=53)	Importance of manure storage/handling (N=20)
Visit farms (N=35)	Employee management/labor issues (N=18)
Use consultants (N=29)	Hire reputable builders/contractors (N=12)
Importance of cash flow/loan availability/financial planning (N=27)	Focus on labor efficiency and profitability (N=12)
Take time/don't hurry/go slow (N=25)	Be open-minded/flexible/willing to change (N=12)
Take advice from farmers/consultants (N=23)	Importance of biosecurity/keeping vaccinations updated (N=10)
Know yourself, your family, your goals (N=20)	Become a people manager rather than a cow manager (N=8)

Table 30 contains the responses producers volunteered when asked to look back and tell about their "best" and "worst" choices. Switching to a milking parlor and freestall facility is obviously the best choice of the majority and their choice of manure handling was the worst of the worst. These comments seem to echo previous results discussed in that producers perceive sand bedding to be superior for cow comfort, but realize it complicates manure handling.

Table 30. Indicate the best and worst expansion choices you made.

Best Expansion Choice	Worst Expansion Choice
Switching to parlor/change to new parlor/efficiency of parlor (N=80)	Manure handling (N=41)
Switching to freestalls/building new freestall barn (N=68)	Not hiring contractor/contractor performance (N=17)
Sand (N=36)	Loans/cost overruns (N=13)
TMR/feeding convenience (N=27)	Disease introduction (N=10)
Employee relations/labor efficiency/working conditions (N=24)	Facility design-curtains, sidewall, ventilation, size, etc. (N=10)
Economics/profitability/cash flow/loans (N=21)	Planning/timing problems (N=8)
Family time/time off (N=19)	Building without future in mind (N=8)

Table 31 contains selected producer comments that were volunteered by producers on the survey form and summarized by common themes.

Table 31. Selected producer comments expressing common themes.

General
➤ You learn more with your ears open and mouth shut.
➤ Be positive always, but also temper that with realism and common sense.
➤ Goal setting and communication amongst partners is critical.

Visiting Other Farms
➤ Take your video camera, notepad, and tape measure. Talk to the people who work on the dairy.
➤ When visiting other operations, ask, "What would you do differently?"

Planning
➤ Plan, plan, plan.
➤ Really go slow. See other setups. Ask how they adapted.
➤ Plan before you break ground. Look to the future. Don't limit your options.

Consultants
➤ Surround yourself with a team of experts and listen to them. Invest money in sound advice.
➤ Don't believe everything a consultant tells you. After it is all done, it is your farm, not theirs, so the decisions need to make sense to you.

Employees
➤ You will be a people not a cow manager.
➤ Listen to your help, they usually have valuable ideas.
➤ Create safe and happy working conditions.
➤ Have systems, SOP (Standard Operating Procedures), employee policies in place.
➤ Take management classes to learn how to manage people.

Economics
➤ Don't cut corners on cost if quality is important.
➤ Figure an additional 25 % cost when budgeting your first year's start-up cost.
➤ Do a financial analysis and long-range planning.
➤ Don't forget about replacements of expansion cattle that you need for 2 years.

Cows
➤ Stay on good vaccination program.
➤ Buy heifers for replacements, not cows.
➤ Cow comfort (treat a cow like you treat yourself).
➤ Cash flow consideration-good cows that are managed well will pay you back.

Facilities
➤ Manure storage and removal is a priority.
➤ Study suitability of site very hard first-don't be afraid to move.
➤ Don't overspend on fancy buildings and equipment. Avoid winter construction.

Quality Management Systems (QMS) as a basis for improvement of milk quality in extension services

Lothar Döring[1] and Hermann H. Swalve[2]

[1] *Milk Recording Agency Saxony-Anhalt, Angerstr. 6, D-06118 Halle, Germany*
[2] *Institute of Animal Breeding and Animal Husbandry, University of Halle, A.-Kuckhoff-Str. 35, D-06099 Halle, Germany*

Summary

The implementation of a quality management system on farms is a basic prerequisite for a safe, sustainable and transparent production and also needed for an optimization of processes on the farms. Since 2000 the Milk Recording Agency (MRA) of Saxony-Anhalt is offering support to farms for a basic quality management system (QMS) specifically designed for farms with different production lines. The certified quality management system comprises the four components animal health, animal welfare, ecological aspects and documentation. For the implementation and operation of the QMS, farmers are supported by production guidelines using clearly defined control points, manuals, registers and forms. The implementation of the QMS on farms in general is only successful with the support of the extension service of the MRA. The extension service includes the analysis of production processes, suggestions for improvement and supplementation of the documentation. Based on the QMS, also complex herd management extension services are offered which are on a new level of extension services and enhance the success of these measurements. This is explained using examples. At present, 178 dairy farms with 76,602 cows in Saxony-Anhalt use the basic quality management guidelines.

Keywords: quality management system, milk quality, extension service

Introduction

The implementation of a quality management system on farms is a basic prerequisite for a safe, sustainable and transparent production and also needed for an optimization of processes on the farms. Since 2000 the Milk Recording Agency (MRA) of Saxony-Anhalt is offering support to farms for a basic quality management system (QMS) specifically designed for farms with different production lines. The certified quality management system comprises the four components animal health, animal welfare, ecological aspects and documentation. For the implementation and operation of the QMS, farmers are supported by production guidelines using clearly defined control points, manuals, registers and forms

Requirements at the production level

Within the entire field of food production, numerous vertical and horizontal ties exist between partners. Every production tier relies on the (raw) products of the supplier who has to fulfil defined standards of products and safety regulations.

All partners within this production process have the task to ensure the quality and safety of their products by means of specific measures. The definition of interfaces between partners and the harmonization necessary are of utmost importance.

Quality management systems (QM-Systems) have proved to be efficient tools to ensure quality and safety within the production process and have gained widespread use in recent years. The use of QM-Systems fits well into the basic requirements for production processes with respect to

- Transparency
- Safety
- Precaution

The importance of quality ensuring measurements has increased in recent years. Reasons for this increase in importance, however, are not only the voluntary efforts of the partners within the production process aimed at ensuring a maximum in product quality. Rather, the chronic nature of quality problems discussed in public and the suspicions mutually raised between market partners have put pressure on the wide distribution of these systems and will continue to do so.

Foodborne diseases have increased with new products and processing methods. Hence, new methods of quality control and well documented methods for back-tracing within all production tiers have gained importance. Furthermore, in many of theses cases, an insufficient compliance with the legislation and a lack of governmental control has become evident.

With this background, politicians and the legislature have recognized the necessity of QM-Systems beyond the legislative regulations as a voluntary measurement since QM-Systems in principle are suited to avoid or reduce quality risks that in short terms are not regulated.

QM-Systems: Aims, possibilities and methods

The implementation of quality management systems serves the following aims:
- Safety measures during the entire production process from start to end
- Protection against unjustified liability claims
- Fulfilment of legal requirements and further demands with respect to the protection of consumers, employees, animals and the environment
- Identification of production risks and their causes
- Liability of third parties
- Proof-enabling documentation

Thus, the implementation of QM-Systems means self-protection and the availability of management aids.

Today, for agricultural enterprises, various QM-Systems are available that differ in:
- The procedure of the implementation
- The organisation and putting-into-practice of the control systems and the accompanying certification which are inherent to the respective system
- The sanctions which are imposed when deviations are identified
- The costs for implementation and operation associated with the system

QM-Systems: Prerequisites

Aimed at a widespread use, QM-Systems for agriculture have to fulfil various prerequisites. Firstly, they have to comply with national and international QM concepts. It is important that they are eligible for official recognition by the European Union. Secondly, QM-Systems should consider the agricultural enterprise in total and the individual parts of the system have to have a simple structure, are to be tailored to the specific enterprise and they should be transparent to the consumer and market partners. It has to be warned against an excess use of criteria and the tools for ensuring quality standards have to have a widespread recognition and acceptance. Finally, the costs for implementation and operation have to be kept as low as possible.

In the year 1998, a base quality management program (BQM-Program) was developed in the state of Saxony-Anhalt which strictly accounts fort these prerequisites. For all important agricultural production lines, standard modules were developed which can be put together to a system covering each individual agricultural enterprise. Interfaces enable the connection of the individual modules.

QM-Systems supported by a milk recording agency: The approach in Saxony-Anhalt

The milk recording agency "LKV Sachsen-Anhalt e.V." plays a key role in the realisation of QM-Systems within the state. The agency is an independent and neutral organisation which acts on behalf of the agricultural enterprises, the dairy plants and the state government in a variety of fields. The main task of the agency is milk recording and milk quality analysis which are the basis for the estimation of breeding values and food safety control. For the estimation of breeding values, yields of individual cows (milk, contents of fat and protein, somatic cell count) are recorded. Optionally, this is supplemented by a recording of milkability. Milk quality analysis pertains to the analysis of raw milk before further processing. The contents of fat and protein, somatic cells, freezing point and the detection of possible inhibitors are recorded.

A further task of the agency is an extension service with respect to all quality aspects of the production of raw milk. Additional services are checks and surveying of sample taking devices in tank trucks, of milk meters and milking equipment in general as well as feed ration planning.

For beef producing farms, recording services comprising individual fattening records (daily gain, fertility, losses, etc.) connected with an extension service based on this data, are offered. Further consulting pertains to the mandatory identification of animals and to the voluntary product labelling.

In the modules pertaining to livestock production the following key points are in the focus of the BQM-Program:
➢ Animal health
➢ Animal welfare
➢ Protection of the environment
➢ Documentation

At all levels of the production process Critical Control Points (CCP) were identified. Different weights are assigned to these CCPs depending on the module and individual farm. Rules for controlling and documentation are derived individually for each farm from the CCPs.

QM-System for raw milk production

Within the raw milk production module the following CCPs, mainly pertaining to animal health and welfare, exist:
1. *Housing and cow comfort*
   e.g. climate, bedding, walking areas, management grouping, feet & legs, treatment, water

2. Feeding

e.g. feed area, feeding places, feeding frequency, grouping, consistence of faeces, milk contents, metabolic parameters, residual feed, feed analysis, ration planning, body condition

3. Milking

e.g. milking parlour, climate, equipment, timing, waiting areas, manner of milking, hygiene, milkability, udder, teats, milk flow

4. Herd replacement with respect to udder health

e.g. buying of replacements, health status, natural service bulls, drying off, calving area, hygiene at calving, control of pathogens, milk for calves, hygiene at parturition, disinfection, udder health data base

5. Protection against infections with mastitis pathogens

e.g. staff hygiene, milking hygiene, disinfection during milking times, housing of diseased animals, teat disinfection, bedding, climate

6. Bacteriological analysis (jointly with udder health service, veterinarian)

e.g. definition and methodology of sampling, checks for first calvers and at parturition in general, drying off, resistance analysis

It is obvious that in raw milk production a large number of CCPs is necessary since a large number of effects exists and especially since combinations of effects are known to have significant influences. Figure 1 gives a brief overview on the effects that are to deal with in raw milk production.

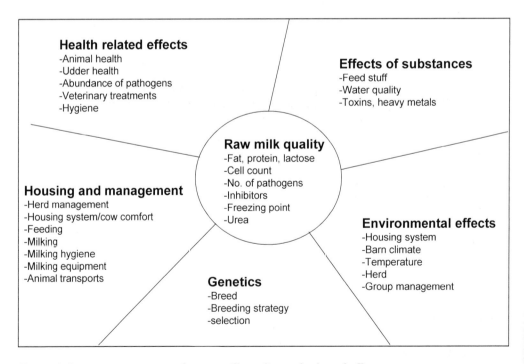

Figure 1. Important parameters for raw milk quality and selected effects.

For ensuring and control of process quality, quality measures are necessary and are to be documented in a transparent and traceable way. The tools needed for this can be differentiated in external and internal tools. External tools, for example, are parameters of the raw milk quality which are determined by the milk recording agency. Internal tools comprise amongst others:

➢ The documentation of all quality relevant work steps
➢ The exact definition of responsibilities and rights
➢ The realization of internal controls and revisions
➢ Continuing education and courses

The documentation of all quality relevant work steps is of utmost importance for production safety and the identification of risks.

One of the most important parameters is the somatic cell count since this variable has a high content of information. The cell count provides information on the health status of the cow and her udder, the quality of milking and milking equipment, housing and cow comfort, adequacy of feeding and herd management in general. Cell count is also a parameter relevant for the product since under German milk quality legislation cell count next to the number of pathogens dominantly determines the milk price.

Increased cell counts in the raw milk lead to a number of economic disadvantages: Reduced milk price, undelivered milk with high cell counts, and costs for acute measures like culling of animals and veterinary treatments. A further reduction in revenues stems from the reduction in yield of diseased cows. This reduction was quantified by Spohr (1998, personal communication) as 10.5 % for a herd cell count of 450,000 cells/ml.

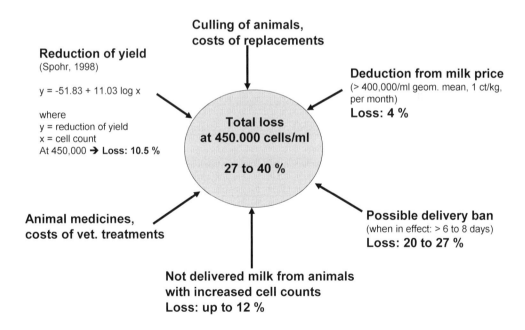

Figure 2. Economical consequences of increased cell counts.

A further advantage of the parameter cell count is the fact that cell counts can be visualized against time in charts. This enables the drawing of quality control charts which are an important tool within the QM-System.

Quality control charts for cell counts contain the expected mean of the cell count, the warning limit and the intervention limit. For the producer, a relation to the legal limit for cell counts should be given, i.e. in Germany 400,000 cells/ml. Therefore, warning and intervention limits in between 300,000 to 400,000 cells/ml are used, depending on the level of the individual farm. If this changes, the limits should be adjusted.

A trend analysis is especially important for an evaluation of the development of cell counts in dairy herds. Using a quality control chart and with or without external help, trends can readily be derived and be used for finding the causes in connection with the CCPs as given in the documentation.

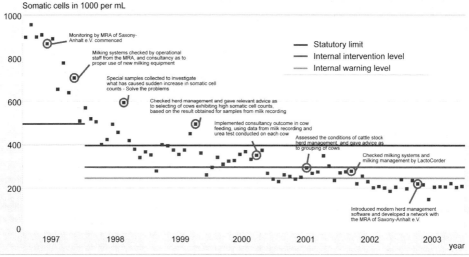

Figure 3. Successful consulting in a farm participating in BQM: Development of cell counts (farm included in 1997 pilot program).

Figure 3 displays a quality control chart of a farm participating in a pilot project for the implementation of a BQM-System. During the given period (1997 to 2003), the following measurements and actions were taken by the extension service of the milk recording agency:

1997: Start of consulting, initial data collection
97/98: Checks of milking equipment and staff courses
1999: Checks of herd management and selection practice
2000: Feed ration planning including individual animal data (urea, fat, protein)
01/01: Checks on milking equipment and milking work (milk flow analysis)
02/03: Implementation of modern herd management software, connection to network of milk recording agency

For the identification of causes and the selection of further parameters to be recorded, the milk recording agency was able to base its work on the extensive documentation done on-

Knowledge transfer in cattle husbandry

farm within the BQM-System. This resulted in a rapid identification of causes and a significant reduction of somatic cell counts.

The cell counts demonstrate how the management of the farm reacted to the trends. Whenever an upward trend was detected from the developments of cell counts and further data, measures were taken. During the observation period, upward trends flatten and the probability of exceeding the limits decreases. The effective use of the chart is also underlined by the fact that no special actions had to be repeated.

Benefits from QM-Systems

By the use of QM-Systems advantages result not only fort he agricultural enterprise but also fort he consulting agency called in for specific cases. An important advantage is the continuing and ongoing work done on farm with respect to the production process and all relevant questions of ensuring quality. The documentation of quality relevant data done routinely helps the agricultural enterprise to do early detection and define risks and possibly call in external help. Consulting then can be more effective and aimed at the specific targets.

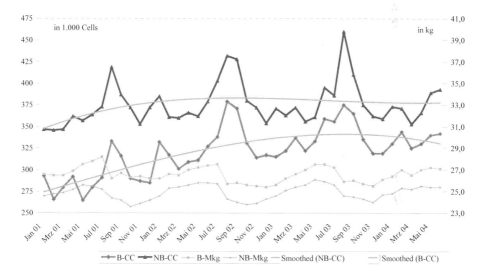

Figure 4. Development of average yields per day and cell counts in BQM- and Non-BQM-farms in Saxony-Anhalt.

Figure 4 displays the development of somatic cell count and milk yield during the period of January 2001 to June 2004 for farms participating in the BQM-Program for raw milk production and non-participating farms. The data comprises 743,511 and 3,563,862 records from individual cows for BQM- and Non-BQM-farms, respectively.

BQM-farms show lower cell counts at all time points compared to Non-BQM-farms. Peaks of cell counts in the summer months are less pronounced in the summer months for BQM-farms. In BQM-farms the critical limit of 400,000 cells/ml throughout the study period is not exceeded. Saxony-Anhalt has suffered from very bad climatic effects in the two most recent years, in 2002 severe floods and in 2003 a severe drought affected the production conditions. The effects of these climatic influences also affect cell counts in a negative way and this applies to both types of farms, the trend is even more visible for BQM-farms. However, since September 2003 this trend has stopped and cell counts have been decreasing continuously in BQM farms.

The profit attributable to an implemented quality ensuring programme is as difficult to quantify as the profit stemming from successful consulting. However, using the approach of Spohr (1998; personal communication; compare Figure 2), the reduction in milk yield of diseased cows can be quantified. For large herd conditions as found in Saxony-Anhalt, this results in an average additional profit for BQM-farms of 13,550 € per year. In Figure 4 it has been shown that average yields for Non-BQM-farms are considerably lower than for farms participating in the program. The difference in yields at least partly may also be attributed to the BQM-program and hence a part of this difference should be added to the benefits although it may be impossible to quantify this sum exactly. For the participating farms, the costs of the BQM-program can be summarized as the yearly fee (1,200 €), the internal costs on farm (especially labour costs), and additional costs for external consulting agencies.

Conclusions

QM-Systems
➢ are offered to farmers in different variants
➢ serve as management tools and as a protection against unjustified liability claims
➢ enable rapid identification of critical points
➢ enable rapid and effective use of external help/consulting
➢ provide optimal and complete data bases at critical control points for consulting/extension agencies
➢ increase in importance for the economic profit of agricultural production

Focused research, information transfer and advice: first evaluation of a new approach undertaken in Emilia-Romagna

Adelfo Magnavacchi[1] and Giancarlo Cargioli[2]

[1] *Centro Ricerche Produzioni Animali CRPA SpA, C. so Garibaldi 42, 42100 Reggio Emilia, Italy*
[2] *Regione Emilia-Romagna, Agricoltura, Viale Silvani 6, 40122 Bologna, Italy*

Summary

High quality products made in an environmentally friendly manner have always been the goal of the agricultural policy of the Region Emilia-Romagna. Parmesan Cheese, Parma Ham, Culatello and a number of other animal products are the testimonials of the success of such a policy. In order to enforce the application of the political guidelines the Regional administration designed a "system for agricultural development" based on private guidance, competition and transparency. The paper illustrates the experience of CRPA SpA, a mixed public/private organisation operating in the animal production sector with the role of collecting the research needs, setting up and managing projects, disseminating the findings of the research. Figures of structural aspects as well as about the effect of «privatising» the traditionally public system and promoting competition, show that strong improvements in efficiency can be reached. Less clear results can be cited for technology transfer and knowledge exploitation where the lack of tools and methods for impact benchmarking does not allow a definitive evaluation. Successful cases and a general hypothesis will then be presented as well as some early trial to measure the efficiency of the information flow and the impact of the new approach at the farm and at the sector level.

Keywords: information transfer, needs, impact of research and extension, practical examples

Figure 1. Emilia-Romagna inside Italy.

Introduction

In the experience being presented in this paper, issues concerning information transfer cannot be afforded without outlining the context in which they happen. In Italy Agriculture matters are mainly a regional competence. Each Region has a government which establishes the agricultural policy, usually fitting within the frame of national and European guidelines. Among the delegated matters are the advisory service, R&D and information. It's up to each Region to invest or not in regional R&D, complementary to the national R&D and regions where agriculture is important usually do. This is the case of Emilia-Romagna (Figure 1), one of the most important Italian agricultural regions. In Figure 2 it is shown that the regional Gross Production Value (GPV) is worth 3,666 M Euro of which 54 % comes from crop production and 46 % from animal production. The latter is composed mainly of milk, poultry, pig and beef. Emilia-Romagna accounts for the highest number of "protected" products among the Italian regions: currently 21 products are recognised by the EU as a PDO or a PGI. Even more important is the market share; 30 % of the total regional agricultural GPV and more than 50 % of the animal production GPV is derived from such "protected" products. Table 1 reports some figures representing the cattle sector.

Table 1. The cattle sector in Emilia-Romagna.

The cattle sector	Units
Dairy farms (n)	7,279
Milk marketed (t)	1,795,500
... of which used for PDO cheese	81 %
Dairies delivering milk for PDO (n)	6,333
Cheese (PDO) factories (n)	575
Estimated Beef farms	4,300
of which of Romagnola beef (breed book)	635

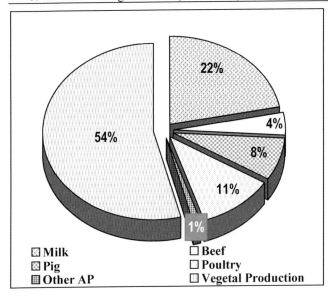

Figure 2: Emilia-Romagna Gross Production Value (GPV) 3,666.9 M € (2002).

In such a scenario, the agricultural policy of the Region Emilia-Romagna has always been geared towards the following strategic objectives:
- keep or bring the greatest part of production on the top shelf of the stores;
- promote the highest possible quality of products as a way to compete in the market;
- sustain the competitiveness of the enterprises in the domestic and external markets by means of "system actions";
- force the production sector to care for environmental sustainability and animal welfare;
- implement the rural development measures, mainly in the mountain part of the territory.

Research, field trials, information and advisory service have been among the most important leverage used for pushing the agro-food sector to pursue those objectives. A first regulation granting funds for R&D was issued in 1988 and after 10 years of application a second regulation has been promulgated and is still in force: the regional law 28/1998. The latter has been designed and discussed under the influence of some criticism about the application of the former: the complaint was that "the research results were of little interest for the enterprises because the research in itself was not really suited to the needs of the production sector". In order to respond to this accusation the Region designed the new law according to a different approach (Figure 3):
- transfer the responsibilities to lead the R&D and the advisory work to the private sector (SMEs and Associations) by means of
  ➲ organising themselves in structures specialised in this matter
  ➲ proposing issues of their interest to work about
  ➲ co-funding the projects
- to fund the "organisation of research activities" and the dissemination of the results
- to publicity call for applications, opening the participation to all the qualified organisations and setting up a transparent evaluation procedure with the support of independent evaluators
- to comply with the EC regulation on State aids for research.

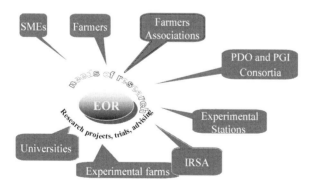

Figure 3. Determination of research needs.

This "approach should have removed any pretext of the production sector for not accepting or not applying research results and should have fostered its willingness to participate in a research activity or in the exploitation of the results". Another expected achievement was an improvement of the efficiency of organisations doing research and similar activities.

Table 2. Results in five years of application of LR 28/98.

	YEAR					
	1999	2000	2001	2002	2003	Total
Applications presented	182	240	256	206	219	1103
Applications approved and funded of which ... [1]	159	176	144	105	84	668
– research project		97	64	36	27	224
– experimental project		15	18	16	27	76
– organisation of research / dissemination results		8	9	9	7	33
– setting up tools supporting the advisory services		29	23	38	18	108
– for advising activity at a regional level		5	10	6	5	26
Millions Euro of direct regional contribution [2]	8.8	10	9.4	9.5	9	46.7
Millions Euro devoted to the provincial projects	6.5	6	6.2	6.2	6.3	31.2

[1] Figures do not include subsequent trances of projects already approved in previous years (see next note)
[2] Because of a national rule on public expenditure, annual amounts include contribution for the corresponding annual trances of projects approved in previous years

After five years of application a first balance can be drawn. From the system performance point of view the results are really positive (Table 2), considering the total managed contribution of about 78 M Euro. More in details, in that period:

- 1,103 proposals (applications) were presented of which 668 have been approved. For a total granted contribution of 46.7 M Euro against a total investment of about 58 M Euro
- 31.2 M Euro proposals were transferred to the Provinces to finance the advisory services delivered at a local level by associations and other private organisations
- the regional service responsible for the regulation has been able to meet all the scheduled deadlines despite the complexity of the rules and of the bureaucracy
- the quality of the projects has noticeably improved
- the agro-food associations and co-operatives have progressively converged in two main "institutions organising research needs" - CRPA and CRPV - operating respectively on the animal and vegetal topics, catching in average 70 % of the projects approved.

From the point of view of the impact of the research results on the agro-food sector data are far less clear. In fact no benchmarks have been set to this end, even because a survey on the existing tools showed that none of them is really suited for small scale application as is usual in a regional territory. An evaluation is being done of the possible use of Dreams, an econometric model and software produced by IFPRI for assessing the effect of the research in a large area of the world. With the lack of tools able to quantify the impact, the Region opted for indicators of performance. One example is the difference in content of somatic cells (Figure 4) and other indicators on farms receiving publicly funded advisors compared with other farms. Such kind of indicators are valuable but rarely definitive because sceptics can always sustain that small differences can be explained by variables different from the one being measured, like the personal attitude of farmers for better management.

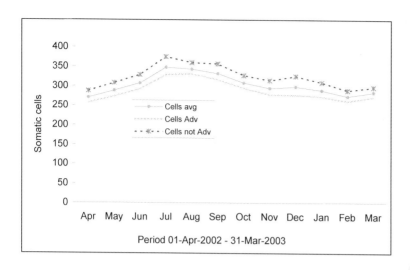

Period 01-Apr-2002 - 31-Mar-2003

Figure 4. Somatic cells comparison in Parmigiano-Reggiano Area.
(Source: AIPLE-Regione Emilia-Romagna)

Table 3. Approved and funded projects for the animal production sectors – (only research projects, 1999-2003).

Topic of research	No.[*]	EURO	%
Feeding	16	1,063,280.81	13.2
Welfare	2	187,758.33	2.3
Economics / management	20	1,064,241.58	13.2
Technology / implements	21	1,266,843.77	15.7
Physiology	19	1,154,766.98	14.3
Environmental aspects	8	679,877.46	8.4
Products quality	28	2,636,734.01	32.7
TOTALS	114	8,053,502.93	100

Animal categories	No.	EURO	%
Ruminants	65	3,781,317.03	47.0
Pigs	25	2,322,553.52	28.8
Poultry and others	24	1,949,632.38	24.2
TOTALS	114	8,053,502.93	100

Products	No.	EURO	%
Milk and derived products	37	2,418,586.41	30.0
Meat and derived products	36	2,976,332.12	37.0
Eggs	2	76,082.26	0.9
Common to all	39	2,582,502.15	32.0
TOTALS	114	8,053,502.93	100

*The number of project refers to the yearly trances of each project, i.e. a 3 year long project counts for 3 projects

New management practices, attitudes and adaptation

Also for organisations such as CRPA, participated and governed by the farmers associations, it is difficult to distinguish between facts and beliefs. Our experience in the arena of the regional research can be drawn in a few figures. Table 3 provides a breakdown of successful applications – the most part presented by CRPA – from which is possible to see the importance of product quality and the equivalent weight of milk and meat products. Coming to information transfer and dissemination actions, figures in Table 4 show the great amount of activities undertaken, provided that it refers only to the direct actions and to ruminants. What is missing is the measure of real leadership or of the transfer of information implemented into farm practices. In order to partially respond to this lack of knowledge CRPA participated in the MARTHA project, funded by the Ministry of Agriculture and leaded by Prof. Giuseppe Pulina of Università di Sassari. MARTHA is a mix of tools and methodologies whose aim is to trace information flow from the origin to the end users. It is based on Internet interface and for testing purpose has been used by 2.600 associates to the national register of Agronomy Doctors. It provides a good view of how and in which degree information generated by research projects reaches the users. Once again the tool seems appropriate but it has not yet been used in an extensive manner for assessing the information transfer for regional projects.

Table 4. Direct communication action of research results for ruminants.

Actions (1999 to 2002)	Units	Users max[1]
Divulgative articles [2]	65	80,000
Technical articles	62	25,000
Seminars, conf., open days	19	50
CRPA Notizie	9	2,500
Books and manuals	6	1,500

[1] Readers of magazines, buyers for books, subscribers for CRPA Notizie average n. of attendants for seminars etc...
[2] Divulgative articles refer to the Agricoltura magazine, edited and distributed by the Region Emilia-Romagna

A final word about the effectiveness of the approach undertaken by Regione Emilia-Romagna from the point of view of end-user satisfaction – that implies an evaluation of the information transfer too. As internal observers it seems to us that things are going better than before and that the involvement of the production sector in the proposition phase - and in some cases in the realisation of the projects - makes them more conscious of the results and of the importance of the work done. However the complaints still exist and the lack of tools able to give a weight to the complains and to the silence of the (satisfied) users is a real threat for the whole system.

In addition to the conclusions expressed on the functioning of the regional approach, a better and more reliable evaluation of the information and technology transfer processes can be probably taken from practical examples, keeping in mind that these are qualitative evaluations and not rigorous statistical investigations. Therefore, some practical examples are presented.

Case 1

Better forage for Parmesan dairy cattle

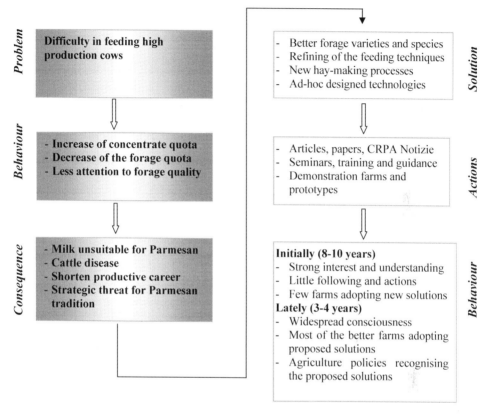

Problem

Difficulty in feeding high production cows

Solution
- Better forage varieties and species
- Refining of the feeding techniques
- New hay-making processes
- Ad-hoc designed technologies

Behaviour

- **Increase of concentrate quota**
- **Decrease of the forage quota**
- **Less attention to forage quality**

Actions
- Articles, papers, CRPA Notizie
- Seminars, training and guidance
- Demonstration farms and prototypes

Consequence

- **Milk unsuitable for Parmesan**
- **Cattle disease**
- **Shorten productive career**
- **Strategic threat for Parmesan tradition**

Behaviour

Initially (8-10 years)
- Strong interest and understanding
- Little following and actions
- Few farms adopting new solutions

Lately (3-4 years)
- Widespread consciousness
- Most of the better farms adopting proposed solutions
- Agriculture policies recognising the proposed solutions

What did not work in the first phase of the information transfer?
- The proposed solutions were more expensive or more complicated than the usual practices
- The proposed solutions were not traditional nor known
- Some unsuccessful adoption of similar technologies were experimented
- CRPA was alone in proposing such solutions
- The suppliers of equipment or seed were not powerful enough to push for the solutions

What triggered the success in the second phase of the information transfer?
- The Parmigiano-Reggiano consortium embraced the proposed solution in philosophy and in practical measures (production rules)
- The public authorities put some aid into improving forage production
- New equipment providers found interest in promoting such technologies
- The long information campaign to spread knowledge and culture

What we learnt
- Science or technical information providers alone are little trusted in agriculture
- Effective information transfer needs strong testimonials, possibly authorities
- Technology transfer needs a commercial power behind it, good ideas alone rarely work
- Uncomfortable or costly innovations take time to be digested, even when they are necessary

Case 2

Good practices in farm building design

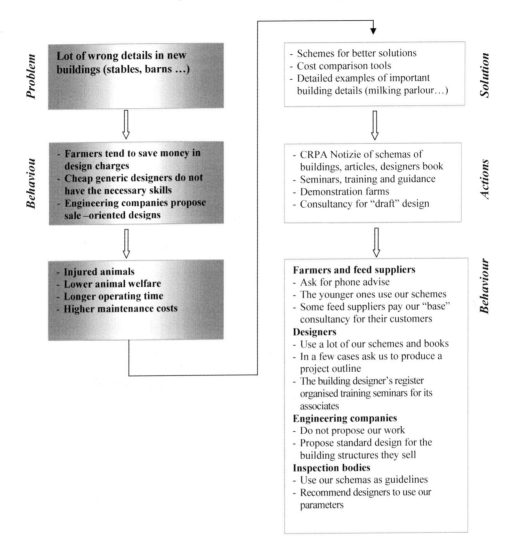

What did work in the information transfer?
- The farmers know the importance of a good building design
- The generic designers want to satisfy customers and CRPA is not a real competitor for them
- The proposed solutions are reasonable, understandable and as comfortable as the alternatives
- The public authorities and inspection bodies find our solutions appropriated

What we learnt
- Once again information transfer get improved if embraced by provider or controller of the farmers. In this situation information becomes as credible as the technology suppliers.
- Scientific results are more credible when related to material aspects (building, machinery …)

Some other interesting examples could be provided, but for exposition purposes it's better to report the conclusions taken from an empirical evaluation of our experience. From the point of view of the **decision making actors**:

- mediation of advisory services is decreasing in importance
- suppliers of equipment and intermediate goods are becoming more and more the advisors and information providers of the farmers
- new alliances between buyers and providers of the farmers for controlling the whole production chain are imposing to the farmers solutions, innovations and prescriptions
- the "political" issues such as food safety, environment, animal welfare, free competition, once transformed in regulations or aids act as guide for innovation implementation
- young and skilled farmers want to be informed themselves of any possible details and are a new target of the information actions

from the point of view of **media and methods:**

- the most effective way to transfer technology, best practices, new processes is to embed their implementation in mandatory or recommended production rules, laws, requisites for aids
- to implement the results effectively using supporting tools such as DSS
- demonstration farms, working prototypes and open days are really effective
- in our experience it is more effective to use a CRPA Notizie, a "vulgarisation" leaflet printed in 4.000 copies, than articles in magazines with some ten thousands subscribers
- seminars, round tables and meetings are still important, but the audience is becoming more and more demanding, so speeches must be short, precise, understandable. The statistics of feedback for our quality system show that attendants are rather severe in voting the quality of the speeches regardless to the reputation of the speaker
- articles are important and in our experience it is better to have a technically complete and understandable article then many less complete articles in lots of magazines
- besides complete articles for farmers and professionals a number of simplified communications must be delivered to get the less technical entourage aware of the innovations and not contrary to them because of lack of knowledge.

The impact of the Zimbabwean change process on dairy farming and farmer attitudes

Stanley Marshall Makuza[1] and Clemens B.A. Wollny[2]

[1] *Department of Animal Science, University of Zimbabwe, P. O. Box MP 167, Mount Pleasant, Harare, Zimbabwe*
[2] *Institute of Animal Breeding and Genetics, University of Göttingen, Kellnerweg 6, 37077, Göttingen, Germany*

Summary

Approximately 95 % of Zimbabwean milk is produced by the large scale commercial sector and only 5 % is contributed by the small scale farmer. Foreign currency shortage is affecting genetic improvement, which depends on imports of genetic material. In the smallholder sector lack of a national breeding programme resulted in indiscriminate crossbreeding. In 1980 Zimbabwe's Land Reform and Resettlement Programme was implemented. In 2000 it was fast tracked for political reasons with the following effects: reduction of the large scale commercial dairy farm sector, destocking of herds, changing of land use and ownership patterns, or relocating to neighbouring countries. Consequently the number of dairy farmers declined by 43 %, the national dairy herd by 63 % and milk deliveries by 50 % since 1999. Currently there are milk and dairy product shortages. The relationship between commercial dairy farmers and the Government is characterised by a negative, hostile and confrontational interaction and attitude. The drastic impact of the ongoing change process requires development of adequate breeding programmes and policies for the future of the dairy industry. Regardless of the legal status the newly resettled Zimbabwean farmers require adequate economic and technical training to improve their skills, knowledge and understanding of dairy farming.

Keywords: land reforms, small scale subsistence farming, large scale commercial farming, change process, attitudes

Introduction

Agriculture is the mainstay of Zimbabwe's economy contributing between 15-20 % of Gross Domestic product (GDP). In a normal rainfall year, agriculture provides 100 % of the nation's food requirements, 60 % of the raw materials for industry, 45 % of foreign exchange earnings and 26 % of formal employment (Country report, 2003). Livestock products contribute about 25 % of value of agricultural output in all farming sectors. Dairy production accounts for 6 % of this value (CSO, 1999). Agriculture grew by an average of 5 % per year during the five years leading to 2000. However, in 2001 it declined by 12.9 % due mainly to the fast tracked Land Reform and Resettlement Programme (LRRP) (Daily News, 2000b). The Holstein, Jersey and crossbred cows are the three major dairy breed groups in Zimbabwe (Makuza, 1995). In the high rainfall regions the Holstein is the predominant dairy cattle breed (Figure 1). The Jersey is the predominant dairy breed in the semi-arid regions, followed by the Holstein, the Red Dane and the dual-purpose Simmental.

Figure 1. Herd with Holstein-Friesian dairy cattle.

Situation analysis

The Dairy Development Programme (DDP) was established in 1983 to promote milk production and processing in the smallholder farming sector. To date, 21 dairy schemes are in operation and 18 milk processing centres have been established throughout the country. The smallholder farmer's contribution to total national milk production is however still small at 5%. Approximately 95 % of Zimbabwean milk is produced by the large scale commercial sector (Country report, 2003). Currently the major constraints to commercial dairying in Zimbabwe are the uncertainty due to the LRRP, foreign currency shortages, high feed costs and price controls coupled with low producer prices.

Local and imported semen is obtained from Animal Breeders Cooperative (ABC), a private breeding company. ABC is the major company in Zimbabwe dealing with semen, embryo and livestock sales, imports and exports. The other service they offer is sale of artificial insemination equipment. However, this company is scaling down operations due to a shrinking client base as a result of the foreign currency shortages. This is affecting genetic improvement, which depends on imports of genetic material mainly from the USA, Canada and the UK. US exports of dairy bull semen to Zimbabwe were worth 294,000 and 133,000 US $ in 1997 and 1998 respectively (US, 1998). We estimate negligible amounts of foreign currency were allocated for dairy semen and embryo imports from 1999 onwards as the political and socio-economic environment worsened. We tried to get figures from ABC with no success.

In the smallholder sector lack of a national breeding programme and policy resulted in indiscriminate crossbreeding and inbreeding. The introduction of exotic dairy breeds in the smallholder farming communities is rapidly substituting the indigenous genetic base (Mhlanga *et al.*, 2002). The genetic impact of this needs to be examined and quantified in order to make adequate recommendations to the Government of Zimbabwe. The drastic impact of the ongoing change process requires development of adequate breeding programmes and policies for the future of the dairy industry. We should point out that the high grade dairy cows are not suitable for the small scale farmer as they depend on availability of external inputs such as veterinary requisites and supplementary feeds (Mhlanga *et al.*, 2002).

The fast tracked LRRP of 2000 has seen 11 million hectares of land compulsorily acquired by the State from about 4000 white commercial farmers (Table 1). It is estimated that about 500-600 white commercial farmers have remained on their farms (Zimbabwe News, 2003a). The goal of the land reform programme in Zimbabwe was the resettlement of people from densely populated communal rural areas to newly acquired farm land. Estimates are that 129,000 families were resettled under fast track A1 model scheme, which is the so called villagised leasehold small scale subsistence farming scheme with each family allocated 5 ha of arable land plus 40-100 ha grazing (Figure 2). About 55,000 families were resettled under fast track A2 indigenous commercial farming scheme (Daily News, 2003a). The A2 commercial scheme is for financially viable individuals given a 99 year leasehold ownership period. Land sizes range from 50 to 400 ha. We must emphasize that there are no title deeds for both the A1 and A2 schemes as it will remain State land. This causes uncertainty and discourages farmers investing in infrastructure and retards agricultural development. The majority of these 'new' farmers are people with few agricultural skills.

Table 1. Resettlement schemes since 2000: more than 11 mio ha were acquired from 4,000 farmers.

Scheme	Ownership	Farm size	No. of families settled
A-1 Subsistence	Leasehold	5 ha arable plus 40 – 100 ha grazing	129,000 (only 50 % up take)
A-2 Commercial	Leasehold (99 years)	50-400 ha	55,000 (only 50 % up take

Figure 2. Small scale subsistence.

Due to the current political situation in Zimbabwe, surveys were not possible to conduct. However, the relationship between commercial dairy farmers (Figure 3) and the Government is characterised by a negative, hostile and confrontational interaction and attitude. Some commercial farmers have formed Justice for Agriculture (JAG), a splinter organisation to the Commercial Farmers Union (CFU), which advocates for legal recourse to the Government's compulsory land acquisitions. JAG criticizes the CFU for being sell-outs, soft and lenient with the Government. Other commercial farmers have now settled in neighbouring countries namely Mozambique, Zambia, South Africa and Botswana (Figure 4).

Figure 3. Commercial farming.

Figure 4. Zimbabwe as part of Africa.

Farm workers have been displaced ballooning the unemployment rate currently estimated to be above 70 % (Zimbabwe News, 2003b). The 'new' farmers disapprove the manner in which Government is carrying out land reform, in particular the lack of clear criteria for allocation of land, violations of one-man-one farm policy and lack of structured support. The

Knowledge transfer in cattle husbandry

'new' settlers still want the Government to supply them with agricultural inputs, equipment, machinery, social service infrastructure and financial credits. This has led to the slow take up of land of about 50 % especially with the A2 model scheme, which has had a negative impact on food security (Daily News, 2003a).

Due mainly to the LRRP, the size of the national dairy herd have declined by 69 % from 190,000 milking cows in 1999 to about 59,200 cows in 2002 due to shooting, looting, indiscriminate slaughter and periodic Foot and Mouth Disease outbreaks due to breakdown of veterinary services (ZDSA, 2003). Assuming a calving rate of 60 % and a sex ratio of 50 : 50, it will take at least 7 years to recover the Zimbabwean dairy herd with AI. From January 2002 to January 2003 an average of 4 dairy producers have ceased operations and 2 new producers registered per month. The implications are a net loss of 2 dairy farmers per month (ZDSA, 2003). Dairy farmers had no practical alternative to maintain sustainable farming. Over the same period, the total number of registered dairy producers declined by 14 % from 324 to 280 and milk processors declined by 44 %. The change process has had a negative effect on milk processing, milk recording and the entire dairy industry infrastructure is different (Table 2). Note that dealers in Table 2 refer to formal processors and marketing channels.

Table 2. Registered formal Zimbabwean dairy dealers and producers from January 2002 to January 2003 (ZDSA, 2003).

Category	January 2002	January 2003	Percent Change
Producer wholesalers	262	235	-10
Producer retailers	34	16	-53
Producer retailer & processors	18	29	+61
Producers who ceased operations	2	5	+150
New producers	3	2	-33
Processors	18	10	-44
Number of registered producers	324	280	-14

The dairy industry was liberalised in 1994 giving rise to informal dairy suppliers and processors which are difficult to quantify. Total milk production declined by 19 % from 1999 to 2002 (Table 3) implying that large and high producing herds have remained in operation (Figure 5). Milk supplies could decline by at least 50 % in 2003 (Daily News, 2003c). Currently there are milk and dairy product shortages. Consequently, dairy products are now a luxury due to limited purchasing power of the majority of the people.

Table 3. Milk produced for sale for the past 4 years in Zimbabwe (ZDSA, 2003).

Year	Volume (Litres)	Annual percentage decline
1999	175 820 810	-
2000	168 374 679	4
2001	163 591 045	3
2002	142 164 963	13

Unless the legal framework is put in place investments in dairying are risky. The newly resettled Zimbabwean farmers require adequate support, economic and technical training to improve their skills, knowledge and understanding of dairy farming. The government has to adopt corrective measures such as viable pricing of milk products and the protection of the few strategic producers and the national dairy herd.

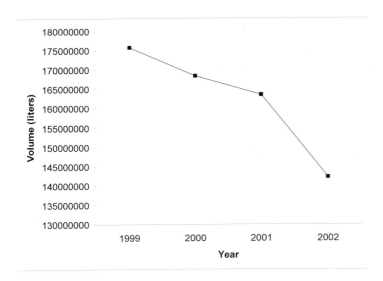

Figure 5. Milk production 1999 to 2002.

Conclusions

The Zimbabwe change process has negatively impacted on the entire dairy sector, dairy farming as large as well as on motivation of farmers trying to manage small scaled dairy farms. The rapid decline of the Dairy Industry shows its vulnerability due to economic, social and political frame conditions. The newly resettled farmers are uncertain about their future in farming.

Figure 6. A way of transporting of milk.

High performing breeds cannot be maintained under the present situation. Future emphasis could be to halt the erosion of the exotic dairy genotypes and for the development of adequate breeding programmes and policies for the future of the dairy industry. Development of a sustainable breeding programme, infrastructure and policy under stabilized frame conditions is one of the future requirements. This includes the need to develop an appropriate dairy breeding scheme for the smallholder farmer as well as the integration of local breeds into a future programme

Acknowledgements

This work was carried out while the first author had a DAAD visiting scholarship for 6 months in the Institute of Animal Breeding and Genetics, Section Animal Breeding and Husbandry in the Tropics and Subtropics, Georg-August University Goettingen in Germany. We like to thank DAAD for this opportunity. The first author would also want to sincerely thank the University of Zimbabwe for granting him sabbatical leave for this period. We also send our gratitude to Mr Nathan Mwale from Zimbabwe Dairy Services Association for providing data as input to this study.

References

CSO - Central Statistical Office, 1999. National Livestock Survey Report, Harare, Zimbabwe. Polycopy.

Daily News, 3 March, 2003a. Harare, Zimbabwe. Website: www.dailynews.co.zw (Accessed 3/3/2003).

Daily News, 13 March, 2003b. Harare, Zimbabwe. Website: www.dailynews.co.zw (Accessed 13/3/2003).

Daily News, 27 March, 2003c. Harare, Zimbabwe. Website: www.dailynews.co.zw (Accessed 27/3/2003).

Makuza, S. M., 1995. Studies on the genetics of dairy cattle in Zimbabwe and North Carolina. PhD dissertation, North Carolina State University, Raleigh, North Carolina, USA.

Mhlanga, F.N., C. T. Khombe & S. M. Makuza, 2002. Indigenous Livestock Genotypes of Zimbabwe. In: Wollny, C. B. A. & N. Demers (editors): Animal genetic resources. Virtual Library. International Livestock Research Institute (ILRI), Addis Ababa, Ethiopia. ISBN 92-9146-124-5, CD-ROM.

Country Report on the State of the World's Animal Genetic Resources, 2003. Ministry of Agriculture, Lands and Rural Resettlement. Zimbabwe, January 2003, Harare, Zimbabwe. 36pp.

USDA - US Department of Commerce, Bureau of Census, 1998. Table 18 of US exports of cattle embryos and semen. Website: www.fas.USDA.gov/dlp2/circular/1998/98-12 T & P/Data/tabl_18.pdf (Accessed 26/6/2003).

ZDSA - Zimbabwe Dairy Services Association, 2003. Regulatory Services Report: January, 2003, Harare, Zimbabwe.

Zimbabwe News, 9 March 2003a. Harare, Zimbabwe.Website: www.zwnews.com (Accessed 9/3/2003).

Zimbabwe News, 24 June 2003b. Harare, Zimbabwe. Website: www.zwnews.com (Accessed 24/6/2003).

Factors determining technology adoption by beef producers in the United States

Larry L. Berger

University of Illinois, 164 Animal Sciences Lab, Urbana, IL. 61801 USA

Summary

The cattle industry in the USA is divided into three distinct segments, dairy production, beef cow-calf production and beef feedlot production. There are 840,000 cow-calf producers, with 78 % of the producers having less than 50 cows. In contrast, there are just over 2000 feedlots that market 87 % of all the finished cattle. Economics drives technology adoption, but the magnitude of improvement has to be much greater for cow-calf producers than feedlots. Ease of technology application is also much more important for cow-calf producers. We developed alkaline peroxide treatment of low quality forages with the cow-calf producer in mind. This treatment process could decrease feed cost by increasing the digestibility of crop residues from 50 % to 70 %, while costing approximately $40 per metric ton. The adoption of this technology has been slow due to low corn prices, and the need for special pumps, scales and mixers which most producers saw as too expensive and complicated. In contrast, we have been working with local feedlots to use ultrasound carcass evaluation as a tool to sort feedlot cattle for niche markets. Despite the fact that the technology cost over $ 20,000 and the return per head is likely $10-20, an increasing number of feedlots are using this method to sort cattle. Because of the large volume, small improvements per head will justify the adoption of this technology.

Keywords: technology adoption, segments of beef industry, economic payback, labor, ease of adoption

Introduction

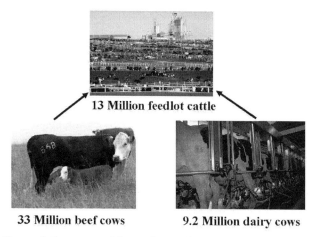

13 Million feedlot cattle

33 Million beef cows   **9.2 Million dairy cows**

Figure 1. Structure of the cattle population in the USA.

The cattle industry in the United States is composed of three independent, but overlapping components, beef cow-calf producers, dairy producers and feedlot operations. As of January 1, 2003 there were approximately 33 million beef cows, 9.2 million dairy cows, 13 million feedlot cattle, and a total cattle population of 96 million head (Figure 1).

Dairy producers focus on milk production (Figure 2) and the majority of the beef generated is in the form of ground beef from cull dairy cows. All male calves are sold at 2-3 days of age and are not typically raised on the dairy farm. These dairy steers, primarily Holsteins, will eventually be finished in feedlots prior to slaughter.

Beef cows (Figure 3) are used to harvest forage, often of poor quality, from land not suitable for crop production. There are approximately 840,000 cow-calf producers in the U.S. with over 75 % of the herds having less than 50 cows. Cow-calf production (Figure 4) is often a secondary enterprise to crop production or non-farm employment. Consequently there is often a low level of management applied to the cow-calf enterprise.

Figure 2. Dairy producer in Illinois.

Figure 3. Beef Cows are used to harvest forage from land not suitable for crops.

In contrast, the U.S. feedlot industry is much more concentrated. There are 780 feedlots with greater than 4,000 herd capacity that market over 80 % of all the feedlot cattle (Figure 5). The feedlot industry is a customer-based enterprise where most of the cattle are not owned by the feedlot. A high level of management is common with wide spread use of nutritional, veterinary and marketing consultants. Feedlots are highly mechanized operations where all inputs are purchased, and throughput is maximized with an average residence time of 150 days.

Figure 4. Characteristics of U.S. Cow-calf production.

Figure 5. Characteristics of U.S. feedlots.

Factors determining technology adoption are similar for cow-calf producers and feedlot operators, but the order of importance may differ. With cow-calf producers, ease of adoption is critically important. For example, less than 5 % of the commercial beef cows (excluding pureblood breeders) are artificially inseminated. Even though this technology has been available for over 50 years, and effective estrus synchronize techniques are available, few cow-calf producers take advantage of artificial insemination. Lack of facilities, technical expertise, and labor are three common reasons given for not adopting this technology. Secondly, the lack of labor is a major factor for cow-calf producers. The wide spread adoption of large round bales is because of the labor saving advantages. Even though storage and feeding losses of large round bales is often between 20 and 40 % of the hay harvested, cow-calf producers utilize this approach because one person can harvest, store and feed large amounts of hay (Figure 6). Economic payback or return on investment is a third consideration for cow-calf producers. Less than 15 % of the cow-calf producers keep detailed records on their cowherd. Obviously, they can't manage what they don't measure and record. However, because the payback is long-term and difficult to quantify, few producers keep good records.

Figure 6. Labor "example".

Figure 7. Labor cost affects technology adoption in the feedlot.

In feedlots economic payback is the driving force behind technology adoption. The use of ultrasound technology to predict carcass merit and then to sort cattle to maximize their value is a good example. Even though the return on investment may only be 2 : 1, many of the more progressive feedlots are examining this technology. Most large feedlots try to operate with one person per 1,000 head of cattle, so labor saving technology is welcomed. For example, covering a large bunker silo with plastic and tires will reduce feed spoilage, but requires large amounts of labor to cover the silo and then to remove the plastic and tires at feeding. We are developing an edible covering that can be sprayed on bunker silos and would eliminate the need for plastic and tires (Figure 7). Even though the initial investment may be greater, feedlots are interested in adopting this technology because of its labor saving potential. Finally, ease of technology adoption is important to feedlot operators, but less of an issue than with cow-calf producers. Steam flaking of grains is a good example (Figure 8). It requires expensive equipment and high levels of management, but because the benefits justify the extra costs and labor, it is a commonly used technology.

Figure 8. Steamed flaked grain (maize).

Conclusion

In the U.S. the successful introduction of new technology to the cattle industry requires an understanding of the factors limiting its adoption. Economic payback, labor, and ease of adoption are the most common factors, but order of importance varies with each segment of the cattle industry.

Use of management practices to differentiate dairy herd environments in Southeastern Sicily

Emiliano Raffrenato[1,2], Pascal Anton Oltenacu[2], Robert W. Blake[2] and Giuseppe Licitra[1,3]

[1] *CoRFiLaC- Consorzio Ricerca Filiera, SP 25 Ragusa – Mare Km 5, Regione Siciliana, 97100 Ragusa, Italy,*
[2] *Department of Animal Science, Cornell University, B21 Morrison Hall 1485, Ithaca, USA,*
[3] *D.A.C.P.A., Universita di Catania, Via Valdisavoja 5, 95123 Catania, Italy*

Summary

Definition of herd environments is nowadays important and useful for research and extension needs purposes. Management practices are the most important determinants of herd environments. Relatively to the geographic area and their association with milk performance and somatic cell concentration, management practices were examined from the survey results. The objectives of this study were: a) to present a methodology based on exogenous variables (practices) to define herd environments; and b) to compare it with an alternative methodology. A survey of 254 farms was undertaken in the summer of 2000. Producers were surveyed to obtain information on nutritional management, udder health, and milking management in addition to other sections about reproductive management, general health, and housing. For Friesian (Brown Swiss) herds there were 10 (11) management practices significantly associated with milk performance and 9 (8) practices with somatic cell concentration. The integrated clustering procedure effectively differentiated the low from the high opportunity environments for both breeds. However, the weights used in the clustering procedure did not have any relevant effect in the differentiation. This approach helps ensure unbiased estimation of phenotypic and genetic parameters and, if information about management inputs is unavailable, herd-year-standard deviation seems to be a reasonable proxy for defining alternative environments.
Keywords: management practices, herd environments, clustering of herds, survey, methodology

Introduction

Many management factors can reduce the profitability of a dairy farm. Furthermore, the extent and quality of management vary considerably among farms, making it difficult for personnel to develop an understanding of the extension needs of the dairy industry within a particular area. A survey was undertaken to provide baseline information on the current state of the management level within this area. Our goal was to identify problem areas related to production to provide direction for planning future extension needs and activities to improve milk production and profits on Hyblean farms. Specific associations of management practices with performance were also analyzed.

Herd performance is greatly influenced by management. Therefore, defining herd environment opportunity based on management practices should be more appropriate but information on herd management practices is seldom available. In this study information regarding herd management practices was obtained directly from the farms via interviews.

Our objectives were:
a) to present a methodology for defining herd environment based on management practices;
b) to compare it with an alternative approach (HYSD).

Materials and methods

A survey of 254 farms enrolled in the *APA* (*Associazione Provinciale Allevatori*), the local dairy recording program, was conducted during the summer of 2000 by the *CoRFiLaC* (*Consorzio Ricerca Filiera Lattiero-Casearia*) of Ragusa to record current management practices. The farms surveyed represented 87 % of the total (297) farms monitored by the *APA*. The questionnaire contained sets of questions about nutrition, udder health, and milking management in addition to other sections about reproductive management, general health, and housing.Data relative to production consisted of mature equivalent milk and somatic cell score (SCS) records for year 1999 from 4114 Friesian and 880 Brown Swiss cows in the 254 herds. A weighted somatic cell score (WSCS) was obtained for each lactation using test day milk yields (m_i) as weighting factors to adjust for stage of lactation:

$$WSCS = \frac{\sum_{i=1}^{n}(m_i SCS_i)}{\sum_{i=1}^{n}m_i}$$

A set of 32 variables reflective of management practices was selected for examination of their association with milk yield and WSCS. Variable selection was based on the biological plausibility of the variable being related to milk production, to somatic cell concentration and as general indicator of management level. None of the selected variables had a correlation coefficient larger than 0.30 and therefore all variables were retained for analysis.

Because of the hierarchically structured data, a multilevel linear model was fitted to study the association of the practices with milk yield and WSCS. The basic multilevel model is:

$$Y_{ij} = \gamma_j + u_j + e_{ij},$$

where
Y_{ij} is the mature equivalent milk record or WSCS for the ith cow in the jth herd,
γ_j is the intercept varying across herds, and both
u_j and e_{ij} are random quantities, whose means are equal to zero.

Season of calving was added to the basic model as a cow level predictor. Season was classified as follows:
➲ November through February was season 1;
➲ March through May was season 2 and
➲ June through October as season 3.
The model now has intercepts and slopes that vary across herds

$$Y_{ij} = \gamma_j + bx_{ij} + u_j + e_{ij}$$

where
Y_{ij} is ME milk or WSCS for the ith cow in the jth herd,
γ_j is the intercept varying across herds
bx_{ij} is season of calving
u_j and e_{ij} are random effects

A forward stepwise variable selection method was used to incorporate the practices at the herd level and to be included in the final model. With this approach the model became conditional on the fixed effects of the practices. The set of variables thus selected was then used to cluster herds into low and high opportunity environments.

Herd environment clustering procedures

Herd management practices Responses to 32 questions were used to cluster herds into contrasting environments. Distance of Jaccard matrices were created with and without weights for the milk associated practices, for WSCS-associated practices, and for their combination. The Lance-Williams (1967) flexible-beta method (Milligan, 1987) was used to cluster herds. Herds in low and high opportunity environment were clustered giving equal weight to practices, and by assigning more weight to practices significantly associated with herd milk production, herd WSCS, or to both outcomes.

Within-herd-year standard deviation (HYSD)

The phenotypic within herd-year standard deviation (HYSD) for 305-day ME milk yield was used to discriminate herds like in other studies (Cienfuegos-Rivas *et al.*, 1999; Costa *et al.*, 2000). Low opportunity environments had HYSD <1260 kg for Friesian herds and HYSD <990 kg for Brown Swiss herds. High opportunity environments were HYSD >=1260 kg for Friesian herds and HYSD >=990 kg for Brown Swiss herds. Accordingly, 91 Friesian herds and 30 Brown Swiss herds were thus allocated to the low HYSD class, and 92 Friesian and 27 Brown Swiss herds were allocated to the high HYSD class.

Results

The management practices selected from the survey and included in the model are among those identified in the literature as most likely associated with milk yield and SCS and most likely indicating the management level of the farm, relatively to the geographic area considered.

Approximately 63 % and 76 % of the variance between herds in the mean milk yield was accounted for by the management practices included in the final model for Friesian and Brown Swiss, respectively. Results from using the WSCS as dependent variable show a slight difference between the two breeds. For Friesian, predictors explained about 62 % of the variance between herds in the mean WSCS, while for Brown Swiss they explained 78 %. This is probably due to the larger number of practices included in the model for the Brown Swiss relative to the Friesian (8 vs. 6). For Friesian herds, there were 10 management practices significantly associated ($P < 0.05$) with milk performance and 9 practices with WSCS. For Brown Swiss herds, there were 11 practices significantly associated ($P < 0.05$) with milk performance and 8 practices with WSCS. Most practices present in the final model had expected positive correlation with milk yield and negative with WSCS for both breeds. Only the presence of pasture in the diet of the animals had an unfavourable association with milk performance and score for Friesian and only with milk for Brown Swiss. Results from Friesian data showed that most of the practices associated with milk yield were those belonging to the feeding management section. For the WSCS practices were from all sections of the screened variables. Relatively to Brown Swiss the practices associated with milk yield and WSCS were in general also from all sections.

All the 32 practices selected were used for the clustering procedures and the significant ones from the above mentioned models were weighted as part of the procedures (see Table 1; example for Friesian herds). All criteria clearly differentiated low from high management environments. Herds in the high opportunity environment had higher mean milk production and lower somatic cell score. When management practices significantly associated with herd performance were weighted, the performance difference between environments was the greatest for both breeds. However, the difference for the HYSD criterion was similar to the other criteria. The average size of the herds confirmed that smaller farms in general provide less privileged environments.

Table 1. High and low opportunity environments by clustering criterion (size, kg of milk, protein and fat, and somatic cell score) for Friesian herds.

Criteria and relative management level/environment	Herds	Size	Milk	Fat	Prot	Score
Clustering without weights[1]						
High	91	56.4	10190.9	326.8	314.6	3.40
Low	92	32.6	8183.6	275.5	249.3	4.12
Clustering with weights for milk[2]						
High	74	60.4	10367.6	330.5	320.3	3.36
Low	109	33.6	8250.3	274.7	251.5	4.06
Clustering with weights for score[3]						
High	114	52.2	9932.7	320.8	306.3	3.50
Low	69	31.7	8044.6	269.9	244.3	4.15
Combined clustering[4]						
High	87	55.1	10149.4	325.8	312.6	3.51
Low	96	34.8	8320.5	279.0	255.1	3.96
Clustering on HYSD[5]						
High	92	52.5	10173.3	328.8	314.0	3.46
Low	91	36.4	8458.4	278.1	258.1	3.93

[1] Unweighted clustering of 32 selected practices.
[2] Weights applied to significant practices from regression on milk.
[3] Weights applied to significant practices from regression on WSCS.
[4] Weights applied to significant practices from both regressions on milk and WSCS.
[5] HYSD classes: low, <1,260 kg; high, >=1,260 kg of milk.

Conclusion

Important management information was obtained through the survey. Not surprisingly, management practices were causally associated with herd performance. Management practices can be effectively utilized to define herd environments.Integrated methodology presented in this study is appropriate for using information from management practices as binary variables to cluster herds into contrasting environments. This approach helps ensure unbiased estimation of phenotypic and genetic parameters. However, if information about management inputs is unavailable, HYSD seems to be a reasonable proxy for defining alternative environments.

Specific extension program is also needed for those farms that do not follow suggested practices. More attention should be toward housing solutions, milking management, and nutrition (TMR). Yet, further study is needed to clarify the pasture effect on milk performance and somatic cell concentration. A reason of the adverse effect found in this study might be related to a high presence of weeds that negatively affect performance. However, pasture is an important economic reality for a low input production system and for smaller farms in general.

References

Raudenbush, S. W., 1993. Hierarchical linear models and experimental design. In: Applied Analysis of Variance in Behavioural Science, Lynne, E. K. (editors), New York: M. Dekker.

Scheibler, D. & W. Schneider, 1985. Monte Carlo tests of the accuracy of cluster analysis algorithms – A comparison of hierarchical and non-hierarchical methods. Multivariate Behavioural Research. 20: 283-304.

Spahr, S. L., 1993. New technologies and decision making in high producing herds. J. Dairy Sci. 76: 3269.

Strategic challenges of Lithuanian cattle breeding sector and attitudes involved

Valdas Dalinkevicius[1], Donata Uchockiene[1], Peter Doubravsky[1], Gintaras Kascenas[2]

[1] *Lithuanian Cattle Breeders Association, Kalvarijos 128, 46005 Kaunas-18, Lithuania*
[2] *"Litgenas", Kalvarijos 128, 46005 Kaunas-18, Lithuania*

Summary

The livestock sector is traditionally the most developed area in Lithuanian agriculture. Cattle breeding in Lithuania is going through a process of reforms at the moment, as well as other agricultural areas. The restructuring of the cattle breeding sector is still ongoing, having problems in all levels because of lack of a suitable strategy. To define and analyze strategic problems of the Lithuanian cattle breeding sector as well as to provide the possible alternatives for their solution, a survey was carried out with 139 respondents from state institutions, private and scientific fields involved in cattle breeding. The questionnaire included 40 questions about cattle breeding services, state support level, cattle breeding strategy in Lithuania, privatization of the breeding sector and legal base. After the calculation of indexes and statistical processing the main problems in Lithuanian cattle breeding were identified. They are the following:
- State support. The main problems are insufficient funding of breeding program preparation and implementation and sub optimal allocation of state funding.
- Privatization of organized cattle breeding. Respondents negatively estimate the fluency of the privatization process and its influence on development of dairy farming.
- Use of additional potential of livestock marketing. The possibilities of livestock marketing are as yet limited in Lithuania.
- Co-operation. Unfortunately, the associations accomplish mainly representative functions and the governmental institutions discharge the main part of breeding work.

Keywords: cattle breeding structures, survey, problems, attitudes, challenges

Introduction

Various legal documents that regulate cattle breeding reorganization projects and cattle breeding development programs were prepared during the past ten years. Active discussions have taken place trying to prepare and implement effective strategy for integration into European Union as well (Figure 1). But cattle breeding specialists recognize that in the sector of cattle breeding some strategic problems exist that will be detailed in this paper. Lack of attention to solve these problems determines that at this moment, Lithuanian cattle breeding does not have a clear vision, priorities, strategy and guidelines for implementation. Hence, the development of the whole sector is halted.

The purpose of this paper is to determine and analyze the strategic problems of cattle breeding in Lithuania and to suggest possible alternatives on how these problems may be solved.

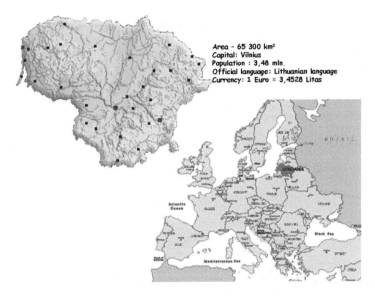

Area - 65 300 km²
Capital: Vilnius
Population : 3,48 mln.
Official language: Lithuanian language
Currency: 1 Euro = 3,4528 Litas

Figure 1. Lithuania as part of Europe.

Strategic problems of the cattle breeding sector

Cattle breeding in Lithuania is in the stage of developing competition at the moment because of the transition of decision making from the state to the private sector, reorganization of technical, legal and other databases preparing to access EU and trying to find an optimal way for developing cattle breeding strategy, on which effective cattle breeding could be created (see Figure 2). Lithuania is continuing reorganization in the sector of cattle breeding and is seeking the right way to develop cattle breeding services, but this process has slowed down because of arising strategic problems. These have to be determined precisely so that solutions can be found.

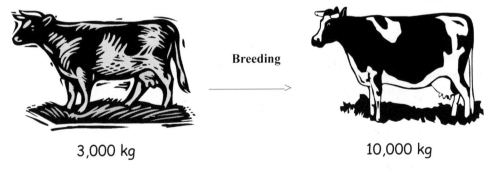

Breeding

3,000 kg   10,000 kg

Figure 2. Breeding goal?

To define and analyze strategic problems of the Lithuanian cattle breeding sector as well as to provide the possible alternatives for their solution, a survey was carried out with 139 respondents from state institutions (36 questionnaires), private (83 questionnaires) and scientific (24 questionnaires) fields involved in cattle breeding. The questionnaire included 40 statements about cattle breeding services; state support level, cattle breeding strategy in Lithuania, privatization of the breeding sector and legal base. The respondents were asked to evaluate these statements in the system 1 to 10 points, where 1 meant "Totally" disagree" and 10 – "Totally agree". After the calculation of indexes and statistical processing the main problems in Lithuanian cattle breeding were identified.

The statistical analysis shows, that the opinion of different groups was the same concerning the questions of selection progress, the activity of associations and their services, privatization of cattle breeding and co-operation. The highest difference was between the opinions of respondents from the state institutions and private field. Respondents from private field (Figure 3) had a much worse opinion about cattle breeding services (0.7 point difference), acquaintance with cattle breeding program (1.6 point difference), legal database (1.5 point difference), usage of additional potential of cattle breeding services (1 point difference) and financial support (1.4 point difference) than respondents from state institutions. The differences of all the five mentioned indexes are statistically significant.

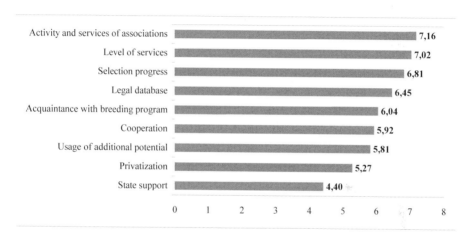

Figure 3. Evaluation of indexes – respondents from the private field.

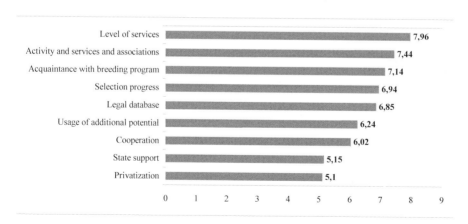

Figure 4. Evaluation of indexes – respondents from the scientific field.

The opinion of respondents from the scientific field (Figure 4) was in the middle between the opinions of the respondents from private field and government institutions. Evaluating the level of cattle breeding services, the acquaintance with the breeding program and using the additional potential of livestock marketing, respondents from scientific field agree with the respondents from the state institutions, but concerning questions of financial support and cattle breeding legal database they agree with the respondents from private field.

Summarizing the results over all 3 groups, the conclusion is that the opinion of all three of them is the same concerning some questions, but in some cases these opinions are different or even opposite. All three groups agree that the level of cattle breeding services is high in Lithuania. All groups give high scores to the activity and services of associations. Respondents from the private field give the highest evaluation to this index, but the respondents from the government institution (Figure 5) describe it as a problematic area. All groups have the same opinion about the selection progress, stating that it is fast enough and the acquaintance with the breeding program, stating that it is widely known. All groups have similar opinion about using the additional potential of livestock marketing, but respondents from the private field see a problem in it. All groups have similar opinion about legal database of cattle breeding, but only respondents from government institutions think, that it is the best developed area of cattle breeding in Lithuania. All groups point out the same problematic areas – state aid, privatization, co-operation. The level of privatization and co-operation is evaluated poorly by all three groups. The respondents from government institutions state that the situation is „very bad" (5.8) concerning state aid, but the opinion of the respondents from the scientific and private field evaluate these areas even worse (5.15 and 4.4 points).

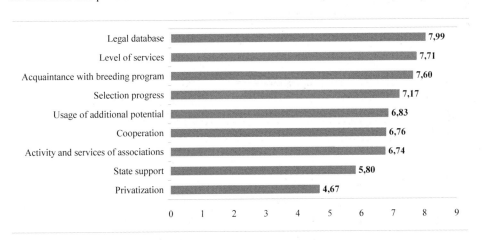

Figure 5. Evaluation of indexes – respondents from government institutions.

Evaluations of different indexes across different groups are often similar, but the rating of indexes and perception of their importance differs in different groups of respondents.

The differences mentioned above lead to the deeper study of these indexes, investigating separate questions and the answers of the different groups to these questions.

Estimation of the state aid was based on summarizing answers to eight questions, including optimal allocation of resources assigned to the support of the cattle breeding sector, sufficiency of the support in general as well as in separate areas of cattle breeding: the control of productivity, breeding programs, scientific research. Average evaluation of the state aid of all the respondents was (4.92 points), the last place among all the indexes. It is the lowest score in the group of private field (4.40 points). In the group of respondents from scientific field and government institutions the score does not reach 6 points (5.15 and 5.80 points). Analyzing answers to different questions, it's evident that respondents of all groups think that government support is insufficient and the allocation of resources is not sufficient. Allocation of resources was given the lowest scores in the group of eight questions. The biggest problems to the respondents from the scientific field seemed the lack of resources and not sufficient sponsorship of scientific research. The support assigned to productivity control, herd book keeping, genetic evaluation was given the highest scores in the question group concerning the level of government support.

The privatization of cattle breeding is one of the slowest among all the sectors in Lithuania. After the research was done, it became evident that privatization processes were given very low scores by the respondents (5.10 points). To evaluate the success of the privatization of cattle breeding, the respondents were given three questions about the influence of privatization on the development of dairy farming, fluency of the privatization process and transition of functions from the state to private sector. All the groups of respondents mentioned the latter as the most problematic area. Transition of functions from the state to private sector was given 5.27 points by the respondents from the private field, 5.10 – from the scientific field, 4.67 – from the government institutions. Respondents from the scientific field and government institutions point out cattle breeding privatization as one of the biggest problems in the sector of cattle breeding nowadays.

Trying to determine potential usage of additional income from cattle breeding evaluation was done in two aspects – estimation of the level of knowledge of cattle breeders and specialists and estimation of breeding cattle market possibilities. Among the three groups the highest potential of additional income in cattle breeding is seen by the respondents from government institutions (6.83 points) and the smallest – by the respondents from the private field (5.81 points). The scientific field representatives are in the middle with the evaluation of 6.24 points. All the respondents emphasize that the level of knowledge differs a lot comparing cattle breeding specialists (7.6 – 8.3 points) and cattle breeders (4.4 – 5.6 points). Estimating the development of cattle breeding market, all the respondents give lower scores to the possibility of selling pedigree cattle (5.1 – 6.7 points). Possibility to buy pedigree cattle is not given high scores as well (6.00 – 6.96 points).

The results of this research allows following conclusions. Cattle breeding privatization was given lowest scores. Respondents criticize both fluency of privatization (5.0 – 5.4 points) and positive influence of privatization to the development of dairy farming (5 –6 points). The transition process of functions from state cattle breeding supervision services to associations is especially criticized. Financial and human resources are still most allocated in the governmental structures. Associations, where functions and responsibilities where transmitted, face the lack of quantity and quality of human and financial resources. Government does not want to give away its functions and associations cannot take them. The same tendencies are evident in the opinion of respondents about the effectiveness of financial support. Respondents point out that state financial support is not sufficient. Giving away essential cattle breeding tasks - that require a lot of resources, from governmental to private sector state does not guarantee sufficient financing of cattle breeding tasks.

Organized cattle breeding and genetic progress is not possible without the development of contribution among separate participants of the cattle breeding system. In the opinion of respondents, individual participants of cattle breeding perform their tasks well and provide cattle breeding services of good quality to cattle breeders, but because the information exchange and contribution is not developed well enough, the final result is much worse than expected, which means that the genetic progress is not on the standard level.

Conclusions

- State support to dairy cattle breeding is not sufficient and the existing resources are not distributed optimally;
- Slow privatisation process is the biggest problem of dairy cattle breeding;
- Co-operation and exchanging information among the participants of separate sectors of breeding is not sufficient;
- Representatives of government institution think that the role of dairy cattle breeding associations is small in developing breeding and dairy products;
- Cattle breeders receive breeding services on high level;
- Respondents from the private field are not acquainted well enough with the program of breeding;
- Research done in the area of dairy cattle in Lithuania is not up to date;
- Respondents from private field and government institutions evaluate different the possibilities of implementing legal acts in Lithuania;
- The level of knowledge of cattle breeders is quite low.

References

Doubravsky, P., 1996. Proposed Criteria for the Development of the most Suitable Cattle Production and Breeding Systems for Lithuania. Bonn: ADT project GmbH.

Esterby–Smith, M., R. Thorpe & A. Lowe, 1997. Management Research: An Introduction (11th reprint). London: Sage.

Food and Agriculture Organization of the United Nations, 1996. Draft Strategy for National Agricultural Development Horizon 2010 – Lithuania. Rome: Author.

Klimas, R., J. Darbutas, B. Zapasnikienė & V. Macijauskienė, 2000. (Nr. 10). Gyvulių veislininkystė ir jos perspektyvos. Žemės ūkis, 22-23.

Lietuvos mokslų akademijos Žemės ūkio ir miškų mokslų skyrius ir Valstybinė gyvulių veislininkystės priežiūros tarnyba prie Žemės ūkio ministerijos, 2000. Gyvulių veislininkystės teoriniai ir praktiniai aspektai integruojantis į Europos Sąjungą. Vilnius: LVA dauginimo priemonių laboratorija.

Lithuanian Institute of Agrarian Economics, 2001. Agriculture in Lithuania: Development and Prospects. Vilnius: Author.

Saikevičius, K., 2000. (Nr. 12). Gyvulių veislininkystės būklė ir perspektyvos Lietuvoje. Žemės ūkis, 37.

Schmidt, F. & H. Momm, 1995. Organization of Competitive Cattle-Breeding Associations in the European Union. Berlin: Blackwell Wissenschafs.

The Future of Livestock Breeding and Service Organizations, 2002. Netherlands: Wageningen Academic Publishers.

Zemeckis, R., 1999. The Main Issues for Agriculture Development of Lithuania.

Design of large scale dairy cattle units in relation to management and animal welfare

Roger W. Palmer

University of Wisconsin-Madison, Dairy Science Department, UW-Extension Dairy Team, 1675 Observatory Drive, Madison, WI 53706-1284, USA

Summary

Selecting the right freestall barn design is a complicated process. The choice of freestall barn design impacts the complete outlay of the dairy; it defines the size of the site needed. The animal management techniques appropriate for each barn type define the amount and type of facilities and equipment in the parlor and treatment areas. For example, if a producer selects the six-row design without self-locking headlocks, then a sort-gate system that selects animals to be bred or treated and a separate area to house and treat these selected animals may be considered. If computer-operated sort gates will be used, then management must ensure that computer-operated sort gates will be used, then management must ensure that computer files are accurate and updated on a timely basis. If the sort gate malfunctions or an animal is not sorted correctly, then procedures must be in place to identify affected cows and to take corrective actions. If palpation rails or other animal batch-handling systems will be used, then the labor requirements of an animal worker must be considered. Overall, the long-term **annual cost** (initial cost plus ongoing labor cost) of the facility and the cost of managing the complete dairy herd should be considered when deciding what type of freestall barn to build. With the high production levels expected of a modern dairy cow, every effort must be made to enhance her comfort. Critically looking at existing facilities and observing freestall usage on different farms should help the producer select the correct stall dimensions, freestall divider type and ventilation. The choice of bedding base and bedding materials should be based partly on the availability and costs associated with each system. There is no ideal combination at this time; each system has both advantages en disadvantages. Most producers in the Midwest USA will choose between stand-based and mattress-based freestalls; both work when managed properly. There are many different types of mattresses available at any given time. Visit farms that have installed the type of bedding that you are considering, and observe its wearability and cow use. The decision of which system to use should be made early in the planning process because it has such a far-reaching impact on manure handling and other facility-related decisions.

Keywords: barn design, handling systems, cow comfort, producer satisfaction, experiences

Introduction

Large scale dairies have become very popular in the past decade because of the adoption of new housing styles and the development of new production-enhancing technologies. These technologies allow producers to enhance labor efficiency, increase profits and improve the quality of life for both dairy owners and workers. The challenge to the managers of these modern large dairy herds is to economically achieve high milk yield without sacrificing animal health and welfare, deterioration of the environment, or human safety.

Dairy producers are basically changing their production systems from where cows are housed and milked in a stall barn to a system characterized by freestall housing, milking in milking parlors and automated feed and manure handling systems.

The dairy manager must evaluate different strategies, then prioritize and implement the changes that will maximize the efficiency of the operation and ensure its long-term viability.

These modern technologies, however, often require larger herds to decrease the investment per animal and better utilize the assets. The best herd size varies, depending on the operator's goals and available resources. Each producer must select and incorporate technologies that allow milk production - now and in the future - at a competitive price. Choose the management system and herd size that best provides a profitable and sustainable business.

For a large dairy to be labor efficient and conducive to animal health and welfare it must be designed to

1) allow groups of animals to be housed with other animals with like needs;
2) allow easy movement of animals;
3) allow easy selection and restraint of animals for health and reproductive treatments;
4) provide for physical separation of disease infected animals; and
5) easy access for feed delivery and manure removal.

The minimum requirements of a good facility design allow far-off dry, close-up dry, maternity, just fresh, milking and sick groups of cows to be housed separately.

The number and size of each group depends on total herd size and the management practices expected. Milking group sizes should be based on expected parlor throughput values which are influenced by milk production levels, milking procedures, milking frequency, etc. Animal health and welfare issues relate directly to profitability in that unhealthy cows lead to high culling rates, low production levels and increased treatment costs. Cows are most susceptible to health problems during the transition phase of their lactation, which starts three weeks before calving to three weeks after calving. To minimize health related problems proper ventilation, freestall design, alley surface design and transition cow facilities must be incorporated into the dairies facility design.

What size and type of milking facility will be used?

To determine the site size needed to support your long-term plans, you need to define the type, size and expected throughput of the milking parlor being considered. The selection of the size and type of parlor, the number of milkers to be used, plus the milking procedures deployed, will determine its capacity. Normally parlors are planned which fully utilize the efforts of a certain number of milkers. Industry standards indicate that one milker can manage a double-8 to double-12 parallel or herringbone parlor, whereas, double-14's to double-24's are considered two-person parlors. Milking shifts of 6.5 hours if 3X milking and 10.5 hours if 2X milking are used to determine the capacity of a parlor and allows time for set-up, clean up, and maintenance of the milking facility.

The throughput of a milking parlor is normally defined in terms of the average number of cows milked per hour (cows/hour) or the number of times the parlor will be filled per hour (turns/hour). The throughput of the milking parlor determines the optimum pen size and maximum herd size for the operation. Current recommendations suggest pens be sized to allow a group of cows to be milked in 60 minutes or less if milked two times per day (2X) or 45 minutes or less if milked three times per day (3X). This rule ensures that cows have sufficient time to eat and lie down, with a reasonable amount of time standing in a holding area, away from feed and water, waiting to be milked. For example, if a system is being designed with a parlor that is expected to have a milking capacity of 72 cows/hour, then pens should be designed to hold 72 cows if 2X milking or 54 cows if 3X milking are planned. Table 1 shows the maximum herd size to expect based on the parlor selected.

Table 1. Herd size potential based on the capacity of different parlor sizes.

	D-8	D-12	D-16	D-24	D-36
Number milkers	1	1	2	2	3
Stalls per side	8	12	16	24	36
Total stalls	16	24	32	48	72
Expected Turns/hour	4.5	4	4.5	4	4.5
Expected Cows/hr	72	96	144	192	324
Total milking cows if 6.5 hr shift, 3X	468	624	936	1248	2106
Total herd with 16 % dry	557	743	1114	1485	2507
Total milking cows if 10.5 hr shift, 2X	756	1008	1512	2016	3402
Total herd with16 % dry	900	1200	1800	2400	4050

Milking parlor considerations

The milking centers, where cows are taken to be milked, vary greatly across the world. Herd size and climate have a bearing on a producer's choice for large herds producers normally select parallel, herringbone, or rotary parlors. For smaller herds and herds making their first move away from milking and stabling their animals in the same barn, producers often choose lower-cost options. Switch milking, flat-barn parlors, swing parlors, used equipment, and use of the existing barns to house the milking center are common. These smaller operations often select parallel, herringbone, or auto-tandem (side opening) parlor types. With all herd sizes, the milking center choice should be based on the short- and long-term goals of the operator and the expected cost to harvest milk.

Switch milking

Switch milking refers to using an existing barn and "switching" pens of animals into the facility to be milked. For instance, a producer with an existing 50-stall barn may build a freestall barn to house cows in pens of 50 cows that can be taken to the old barn and milked. This option has a low capital cost, because existing equipment and facilities are used, but the labor cost associated with milking and moving cows is often high. Walking distances of over 500 feet have been used successfully in the past. Cows adapt well to the walk, but proper lighting, snow removal, and other worker comfort issues must be considered.

Flat-barn parlors

Flat-barn parlors are normally built in existing buildings and are similar to switch milking in that cows are brought from a freestall barn or pasture to an existing barn to be milked. Rather than using all 50 stalls, some of the stalls would be used to milk cows and other stalls would be removed to provide space for a holding area. For instance, eight stalls on each side of the barn could be left and four milker units used on each side. These eight units (4 per side) would be used to milk half of the cows locked in the 16 stalls. When the first group of cows has been milked the units would be switched to the other eight cows and the just-milked cows removed from the barn, and eight more cows brought from the holding area to be ready for the next shift of the milk machines. These flat barns are normally more labor efficient than switch milking in existing barns since the milking machine units do not need to be moved.

Swing parlors

The term swing parlor refers to a parlor in which one set of milking machines is installed and used to milk both sides of a pit parlor. A double-eight parlor would have eight milking machine units. Eight cows on one side of the pit would be milked first and then the units swung to milk eight cows on the other side. This is the parlor of choice for producers who want to milk cows fast with a low investment per animal. Parlors are normally of the herringbone type with varying cluster spacing depending on the animal angle selected for positioning the cows. These parlors are often built with a simple inexpensive rump rail to restrain animals. To keep construction costs low, this rump rail often does not have manure splashguards, the platform floor does not have gutter grates, etc. to reduce manure splattering. Therefore, these parlors are often considered a dirtier place to work than conventional parlors. Milk lines are normally mounted above the milker units (referred to as a high-line), which may have an impact on vacuum levels and increase the incidence of mastitis. Since units from one side must be taken to the other side, a slow milking cow on one side may cause the cow on the other side to have her unit attached after her milk let-down response.

Herringbone pit parlors

Herringbone parlors get their name from the angle at which cows are positioned with respect to the pit wall. Normally a row of cows is positioned at a 45-degree angle on each side of the parlor pit. Milker units are applied to the udder from the side of the cow and this configuration allows an arm type take-off to be used. With American Holstein cows, herringbone parlors require about 45" of parlor pit length per stall. This is less than auto-tandem and more than parallel parlors and should be taken into account when building a large parlor. Since milking machines are applied from the cow's side, the animal can kick the operator easier than she can in some other parlor types. Gutters and grates are normally used with herringbone parlors to catch manure and urine and to minimize manure splattering on the operator.

Parallel pit parlors

With parallel parlors, cows stand perpendicular to the pit wall and milking machines are applied between the hind legs of the animal. Parallel stalls require only about 27" of pit length per stall and are normally chosen for large parlors to minimize walking distance. Parallel parlors are normally considered safer because stalls are built with a rear rail that prevents cows from kicking back and hitting the operator. Since the milking machine must be attached between the legs of the animal, the operator must be careful to remove any manure from the milking machine claw to prevent contamination of the milk. No arm take-offs has been designed for this parlor type, so a rope or chain type take-off must be used. To minimize manure splattering, these parlors often have butt pans mounted on the rear of the stall to catch manure or urine generated.

With both parallel and herringbone parlors cows are loaded, milked, and released in batches. For instance, with a double-8, eight cows are loaded on a side, prepped, milked and then released. This batch handling of cows can support a wide range of efficient milking procedures.

Auto-tandem pit parlors

Auto-tandem parlors are often referred to as side-openers, because cows enter the milk stall from the side, off a lane that runs behind each row of stalls. With this parlor type cows stand in a line and the length of each stall must be as long as the cow being milked. Cows are loaded individually and are allowed to exit at random times. This is different than the parallel or herringbone parlors, which milk cows in groups. Since cows can exit when they are finished (i.e. they don't need to wait till the last cow in the group is finished), the number of cows milked per stall per hour is higher than with parallel or herringbone parlors. But because the stalls are longer, large parlors of this type are not recommended, which limits the maximum herd size milked with this type of equipment. Since cows enter and exit at random times, it is difficult to efficiently implement some recommended milking procedures.

Rotary pit parlors

Rotary parlors are built so cows enter from a fixed point, ride a rotating disk, and exit at a second fixed point. This technology is not new but has been regaining popularity lately with large operations. Cows seem to enjoy the ride and often fight to get on the rotating disk. Since the disk rotates, operators remain at fixed positions and cows pass by as the disk turns. One operator is usually required for each function performed and training of milk machine operators is simplified because each employee has only one function to perform. After the milking unit is attached, the cow moves past the operator who attached the claw. If a milking machine liner slips or a milking machine claw falls off, a worker must leave his position and go to the problem machine to adjust or reattach it. The capacity of rotary parlors is limited by the time it takes to load each animal onto the disk. If a 10-second average load time is achieved, a theoretical throughput of 360 cows per hour could be expected. Since not every stall will get filled as the disk turns, some cows will require a second trip around in order to be completely milked and since the disk must be stopped in case of problems, it will never reach this theoretical limit. Smith et al. recommend sizing these parlors assuming 11-12 seconds per stall to load cows with 80 % of the theoretical throughput expected.

Rotary parlors have a reputation for milking large numbers of cows per hour if a minimal milking procedure is used. Research by Smith et al. shows their throughput, on a cows-per-worker-hour basis, is less than other parlor types when a full-prep milking procedure is used, because of the number of workers needed. Rotary parlors usually require a larger building than herringbone or parallel parlors with the same milking capacity and have higher equipment costs. Furthermore, expansion of the parlor is difficult. It should also be noted that if an operator leaves a pit, a replacement worker is needed or the parlor must be stopped. Since operators continuously perform the same repetitive task, workers are often rotated to avoid operator boredom.

Robotic milkers

The newest way to milk cows is with a robotic milker. They have been developed in Europe and been used on several hundred farms for several years. They are currently being tested and sold, on a limited basis, in the U.S. and Canada. They hold a lot of promise for traditional producers who wish to increase herd size and milking frequency without adding hired labor.

Price, milk quality issues, and current milk marketing regulations need to be resolved at this time.

How will the parlor complex and freestall barns be arranged?

In the past there have been two common ways to arrange freestall barns and parlor/holding area complexes. The H-style configuration has two freestall barns with a parlor/holding area complex between the two barns. The T-style configuration has the parlor/holding area complex perpendicular to and attached to one of the freestall barns, with the second freestall barn behind and parallel to the first barn. Lately a third complex type referred to as the Modified-H-style has become very popular. It is has the parlor complex arranged parallel to both barns like the H-style, but the parlor complex is located beside the barns rather than between them.

The advantages of the **H-style** arrangement:
1. Space is available behind the parlor/holding area for the special-needs barn. This is a convenient location for sick and fresh cow. Pens in this barn are closest to the parlor so these cows have less distance to walk. It is centrally located to both freestall barns so employees can easily monitor cows, and work activities are near the office and parlor areas.
2. The orientation of holding area is the same as the freestall barns, providing proper natural ventilation to both.
3. Cow flow from either freestall barn does not interfere with the other freestall barn, which allows more time for non-milking activities (feeding, bedding, etc.) to be performed between milkings for each barn.
4. Often the average walking distance for groups of cows is shorter.

The advantages of the **T-style** arrangement:
1. Manure from the holding area can be easily incorporated into the manure handling of the attached freestall barn.
2. Only one connector barn or alley is needed.

The advantage of the **Modified-T-style** arrangement:
1. Space is available behind the parlor/holding area for the special-needs barn. This is a convenient location for sick and fresh cow. Pens in this barn are closest to the parlor so these cows have less distance to walk. Work activities are near the office and parlor areas.
2. The orientation of holding area is the same as the freestall barns, providing proper natural ventilation to both.
3. Since the special-needs barn is offset from the parlor/holding area complex, the drive-through feed alley for the special-needs barn can be easily accessed.
4. Since all cows must enter and leave the parlor complex via the same connector barn it allows one site to access all cows going to or returning from the parlor. This can decrease the number of and cost of implementing palpation rails, sort gates, animal oilers, etc.

If the T-style configuration is selected, some portion of the two freestall barns may be used to house special-needs animals; a special-needs barn can be attached to the parlor building; or a separate barn can be constructed near the parlor/holding area/freestall barn complex. When choosing between these alternatives labor efficiency and operator convenience should be considered. The ability to easily monitor, move and treat animals with special needs is very important.

Freestall barn types

Most freestall barns are built with rows of freestalls parallel to the feed manager to allow manure to be easily moved to the center or end of the barn. Barn designs with 1, 2, 3, 4, 5 or 6 rows of stalls are common. Increasing the number of rows of stalls in a barn increases its width and increases the difficulty to ventilate. Barns with one to three rows of cows placed on one side of a feed alley are narrow, easily ventilated, and inexpensive because of the rafter length, but only one side of animals uses the feed delivery lane. By placing rows of freestalls on both sides of the feed delivery lane it is better utilized, but the barn then becomes wider, higher, more costly, and harder to ventilate. The number of rows of cows and the freestall width selected determines the amount of feed space provided per cow along the feed manger (i.e., three rows of 48" stall provides about 16" of feed space per animal). Since barns with one row of stalls would provide more feed space than recommended, they often have feed space on only a portion of their length and the additional length, is used for additional freestalls.

Most barns currently being built are of the 2-row or 3-row design with drive-by feeding or the 4-row or 6-row with drive-through feeding. The drive-through freestall barns normally are constructed to house four pens of cows. This aids cow flow because each pen of cows can exit the center of the barn to be milked and return without interfering with any other cows. Since four pens of cows use a common lane to access the parlor, it is practical to cover this lane and protect the animals and workers from the weather. It also provides an efficient way to consolidate the manure removed from each of the four pens at one common "center alley."

Choosing between 4-row and 6-row designs

Often a manager needs to select between two different alternatives when a new facility is built. One classic example is the choice between a 6-row barn (3 rows of stalls on each side of a drive through feed manger) and a 4-row barn (2 rows of stalls on each side). This is a critical decision point for the producer making a buying decision because of the 10-20 year useful lifetime of the barn. Remember the points mentioned below are similar for the 2-row versus 3-row decision.

When making the decision on which type of barn to build the following three primary factors (the 3'C's of freestall barns) must be considered: **Cost, Cow Comfort and Convenience of Animal Handling.** In some cases secondary considerations, such as site size restrictions also need to be considered.

Table 2. Capital cost per cow differences for 4- and 6-row barns considering recommended stocking rates.

	4-Row	6-Row	Difference
Number Cows	144	144	
Stocking Rate	112.5%	100%	
Number Stalls	128	144	
Cost / Stall	$1200	$1000	$200 (+20 %)
Total Cost	$153,000	$144,000	
Cost / Cow	$1067	$1000	$67 (+6.5 %)

When estimating the **cost** of the freestall barn, consider the stocking rate and evaluate the price on a cost per cow rather than a cost per stall basis. 6-row barns normally are $100 and

$200/stall cheaper, but higher stocking rates are recommended for 4-row barns because of the feed space and animal density issues. Table 2 illustrates the effect on cost differences per cow when overstocking is considered. This example shows the dramatic effect of stocking rate on the decision process in that the 20 % higher cost of a 4-row barn on a per stall basis is only 6.5% higher on a per cow basis.

The second major area of consideration is **cow comfort**. The term cow comfort has become the theme of the industry as we increase the expectations of our dairy herds. Many herds in our country currently have rolling herd averages in excess of 30,000 pounds of milk per year and daily tank averages over 100 pounds per day. These production values can only be achieved by providing an environment conducive to high production. Four-row barns have two rows of cows accessing two manure alleys whereas 6-row barns have three rows of cows on the same two manure alleys. This design difference increases animal density in a 6-row barn and decreases the animals' ability to get to feed and water. This congestion factor coupled with the decreased feed space provided by the 6-row option, as shown in Table 3, needs to be considered when evaluating the relative cow comfort factors of each. 6-row barns are also wider, making them more difficult to ventilate. Wider barns with the same sidewall height are taller and need to be separated more from other buildings to ensure proper airflow, making the overall complex shorter and wider. The impact of this shorter and wider footprint of the 6-row design on available sites should be considered.

Table 3. A comparison of factors relating to cow comfort of 4-row and 6-row barns of equal capacity.

	4-Row	6-Row	6-Row Difference
Barn Length	144 ft	105 ft	39 ft Shorter
Barn Width	92 ft	106 ft	14 ft Wider
Barn Space /cow	115 square ft	87 square ft	24 % less Barn Space
Barn Air Space /cow	1979 cubic ft	1644 cubic ft	17 % less Air Space
Feed Space /cow	24 inches	17.5 inches	27 % less Feed Space

The third factor, **convenience of animal handling**, is probably the most important question to address because any inefficiency built into the facilities will impact the business's bottom line for a long time. How you plan to sort, restrain, and treat cows for all the herd's health and reproductive needs directly relates to the dairies labor requirements and cost. The 4-row design lends itself to self-locking manger stalls because it has 24" of feed space per animal, whereas, self-locking manger stalls are not recommended in a 6-row barn because they only provide 16-18" of manger space, which reduces the number of animals that can eat at any one time. If self-locks are installed in 6-row barns (or 4-row barns with extremely high stocking levels), dry matter intake may be depressed because of the reduced number of animals that can eat at the bunk and result in less production per cow. With the 6-row choice, a separate area is often provided to treat and breed animals away from the freestall barn. Any additional costs associated with this area should be considered when comparing the cost per cow of each option. The task of separating animals, treating them in a different area, and then returning them to their pen often adds expense in the form of additional labor and must also be considered.

What production differences have been reported?

The 1999-Wisconsin Dairy Modernization Project (Bewley *et al.*, 2001) surveyed producers who had expanded their herd size by at least 40 % from 1994 to 1998. Table 4 shows the DHI Rolling Herd average values for those farms that build new 4-row or 6-row freestall barns with drive-through feeding. Production was not significantly different in 1994, but was in 1998. Producers who selected the 4-row option had higher production in 1994 and this production advantage increased by almost 600 pounds by 1998, resulting in more than 1900 pounds more milk per cow per year.

Table 4. DHI Rolling Herd Average milk production values of dairy herds which built 4-row and 6-row barns when they expanded herd size.

	Number Herds	'94 RHA Milk	After Expansion	Change
4-Row	53	21,669	23,644[a]	+1974
6-Row	42	20,351	21,733[b]	+1382
Difference		+1318	+1911	+592

[a, b] Percentages with different superscripts differ ($P < .05$).

4-Row Head-to-Head versus Tail-to-Tail?

If you are considering building a 4-row barn, several factors should be considered when deciding if a head-to-head or a tail-to-tail configuration should be selected. The following is a list of characteristics of 4-row head-to-head configuration barns. Much of the logic can be extended to other barn types that have similar characteristics.
No freestalls on outside wall
- Adv-Better air flow
- Adv-Easier sidewall construction, curtain protection, etc.
- Adv-Cow protected from sun and rain without need of roof overhang
- Dis- Cold weather manure removal
- Dis- Fewer stalls in same length barn

Cows lunge into adjacent stall
- Adv-Need ½' less barn width per row of stalls (15' vs 16' each side)
- Adv-Wider center crossover allows longer water tank
- Dis- Cows' heads near each other, summer heat concern?

Cows can access freestalls and feed manger from common alley
- Adv-Allows cows choice if locked onto feed manger side
- Dis- Cows can't be locked away from stalls
- Dis- Manure alley on manger sides may need to be wider for stall access?

Each row of stalls accessed from different manure alley
- Adv-Two routes between feeding and resting areas
- Adv-Cows don't interfere with each other as they exit stalls?
- Dis- Bedding of stalls may need to be done from two manure alley

Characteristics of a 4-row tail-to-tail configuration to considered

The following is a list of characteristics of 4-row tail-to-tail configuration barns. The advantages and disadvantages pertain to other barn types that have similar characteristics.
Freestalls along outside wall
- Adv-Manure farther from outside wall, less freezing in cold climates
- Adv-More stalls in same length barn
- Dis- Cows on outside wall obstruct air flow, may need taller sidewalls to compensate
- Dis- Cost of barn roof overhang needed to protect outside row of stalls from sun and rain
- Dis- More difficult sidewall construction, curtain protection, etc.

Cows can't lunge into adjacent stall
- Adv-Cows' heads not near each other, less summer heat concern?
- Dis- Cows can't lunge into adjacent stall, need ½' more barn width per row

Cows can't access freetalls and feed manger from common alley
- Adv-Cows can be isolated at feed manger or in freestalls
- Adv-Cows don't need to access stalls from manger side manure alley, narrower alley?
- Dis- Cows have no choice to eat or lie down if locked onto one alley

Center row of stalls narrower, smaller crossover area
- Adv-Less space to hand scrape
- Dis- Less space for water tank

Each row of stalls accessed from the same manure alley
- Adv-Bedding of stalls may be done from one manure alley, less labor and doors?
- Dis- More manure in stall alley than manger side, may complicate manure removal
- Dis- One route between feeding and resting areas, more cow congestion?
- Dis- Cows may interfere with each other as they exit stalls

How and where will animals be handled?

All animals must periodically be isolated and restrained for physical examinations, vaccinations, artificial inseminations, pregnancy checks, treatment, dehorning, calving, etc. on a regular basis. The animal flow, or paths followed by the cow as she is moved to the parlor, to her home pen, to a different location for treatment and any location changes needed throughout the lactation should be considered. The facility design and equipment selected influences work routines, labor requirements and animal stress levels associated with each of these activities.

Cattle's fear of people can be a major source of stress. Stressed animals produce less milk and milking efficiency is reduced. Stressed cattle are difficult to handle and increase the risk of accidents and injuries for handlers and animals. Poorly designed handling facilities cause animals to balk. Properly designed handling facilities lead to easier flow of animals, reduce the need for rough handling, and result in tamer, less fearful animals.

When planning treatment facilities the following recommendations should be considered:
- One person should be able to isolate and restrain an animal safely and conveniently.
- Components should be selected and constructed to reduce the possibility of injury to operators and animals
- Construction should withstand the abuse by 1,500 pound cows and by equipment used to clean the area

- Access to running water, medical supplies, records, and parking for the veterinarian and hoof trimmer's service vehicles should be included
- Good lighting should be provided
- Detailed drawing of how gates will be used to form a funnel to direct reluctant animals into a stanchion or lockup and how people will access restrained animals when performing needed activities should be thoroughly reviewed
- Animal handling systems designers should consider their impact on left- and right-handed veterinarians or managers
- A heated storage area for supplies and equipment near the animal treatment area should be provided
- Man passes and bypass lanes near animal treatment areas to support animal movement should be provided
- If animals will be calved on-site, the maternity location should be situated so animals are observed easily and often
- Sick cows should be housed separately from the maternity area
- If dry cows are be housed at a remote location, the facilities should be designed to support their separation and movements.

Animal handling - possible systems

Most new parlor/freestall operations fall into one of two different types of systems. The "Animal Management Activities" of sorting, restraining, and treating are often done in the freestall unit where the animals are housed (home-based) or in some special area away from where they are normally housed (treatment-area-based).

Home-based systems normally utilize self-locking manger stalls, where cows lock themselves in place upon returning to a manger full of fresh feed after being milked. The self-locking feature is activated when the animal puts her head in a stanchion to eat. A few dairy producers treat animals by cornering them in a freestall. This practice is discouraged because of the safety concerns if the animal moves or the worker slips.

Treatment-area-based systems use sort gates to separate selected animals from their group as they leave the milking parlor. These sort gates can be manually controlled by the parlor operator or controlled automatically by a computer if animals are electronically identified. Animals sorted in this manner may be directed into a palpation rail (also referred to as a management rail system) system or placed in a holding pen and handled using a head chute or some other restraint system.

Home-based animal handling systems

Dairy managers who select the home-based system must evaluate the cost of the self-locking manger stalls (about $60 per head lock) versus the cost of a separate treatment area, plus any labor savings over time. Producers report the following advantages of the self-locking manger stalls:
- Less traumatic handling of cows since they are treated in familiar surroundings
- Cows may eat their proper ration while waiting to be treated
- No time is wasted returning animals to their lot after treatment, because they were restrained in their own pen
- Manger uprights minimize the effect of boss-cows dominating a large section of the feed bunk and can decrease feed wastage

- Large numbers of cows can be automatically restrained saving labor for routine tasks such as tail chalking, hoof spraying, bST injections, etc.
- Manure from restrained animals is handled with normal procedures and should be incorporated into the planned manure handling system
- Locking cows after milking allows teat sphincter muscles to close before the cow lies down, thereby decreasing the possibility of mastitis
- Parlor efficiency can be improved because flow of cows leaving the parlor does not need to be channeled through a narrow sort lane; and operator time to sort or move animals is avoided.

Users of self-locking manger stalls sometimes express concerns about the extra noise generated by some brands and difficulty in finding a specific animal when the cows are caught and restrained in a random order.

Some producers have installed self-locks in only a portion of each housing area that can be gated off and used to treat groups of animals moved from a sort area. This technique can decrease the initial expenditure for self-locks but may complicate animal management since animals will probably show a preference for eating in the section containing no self-locks.

Treatment-area-based animal handling systems

With treatment-area-based systems, animals are sorted and taken to a special place for restraint and treatment. With this type system, the manager must be concerned with the length of time the animal will be away from her home pen and how she will be returned. Labor requirements, availability of feed and water, the effects of the additional stress placed on animals, plus handling of manure are some of the issues to consider when making this choice.

With treatment-area-based systems, cows are often sorted as they leave the milking parlor. Cows need to be diverted through a narrow alley that allows them to be identified and diverted to a catch lane, catch pen, or palpation rail. This animal selection process can be installed anywhere in the path as the animal returns home. If sorting is done manually, it should be done near the rear of the parlor to be easily viewed by the operator, but if automatic sorting is used, it should be located near the end of the return lane to improve cow movement. Maintaining a smooth flow of animals, walking slowly though an automatic sort gate, generally will improve a sort gates ability to accurately select desired animals. Animals that bolt through a sort gate can be selected or missed depending on the position of the sort gate.

Once cows are sorted, they can be restrained and treated using a palpation rail or a chute located in the catch lane, or then may be taken to a catch pen containing self-locking stalls. One major consideration with treatment-area-based systems is that animals returning to their home pen after being treated may use the same traffic lanes as animals being milked. This can cause delays and additional labor to move gates, etc. to prevent mixing of groups.

Palpation rail systems

Palpation rails, also referred to as management rails, are simple structures that allow a group of animals to be restrained in a common area. Front pipes position animal heads and a rail in the rear restrains the animals but is low enough to allow animal to be palpated. Animals are positioned and restrained in a herringbone fashion as they are given rectal examinations, are bred, or given shots.

An advantage of palpation rail systems is that they can be placed in normal cow flow areas, simplifying the selection and restraining of animals. This can be a disadvantage in that animals must be treated promptly to provide space for subsequent animals or else a build up of animals waiting to be treated can disrupt animal flow. It also can necessitate the veterinarian or herdsperson being on site for an extended amount of time waiting for all animals to be selected.

Palpation rails are often considered as an alternative to self-locking manger stalls because the cost per animal is less for large herds and they offer safer working conditions since they prevent animals from kicking the worker behind the animal.

Separate treatment areas

As herd size increases, the need for a special area where cows can be taken and worked on increases. This treatment area should be conveniently located, have a head chute and medical supplies, and be well lighted. This special treatment area is often located near the maternity and sick cow areas so facilities and equipment can be shared and workers engaged in other activities can monitor cows. When these areas are combined they are often referred to as special needs barns.

One or more holding pens should be designed to support the needs of the dairy. If animals are selected from several different milk groups and placed in a common pen, they will need to be sorted later, which increases labor inputs. If animals are to remain in these pens for a substantial amount of time waiting to be treated, feed and water should be provided.

This special treatment area can be smaller for herds that do most of their animal handling in palpation rails or self-locks, since it will be used mainly for major surgeries, hoof trimming, etc. For herds not selecting one of these animal-handling systems, this area should be larger, and keeping animals from different groups separate should be considered.

Electronic sort gates

For many large herds, electronic sort gates are used that automatically identify and sort animals that need special attention. With these types of systems animals are identified by an electronic transponder that is temporarily or permanently placed on the animal.

Temporary identification is normally attached to the leg of the animal during milking and causes the animal to be selected as she returns from the parlor. This is a less expensive system but requires extra labor in the parlor, so it may slow the milking process and is governed by the milker's ability to identify cows needing attention.

The permanent identification system is more expensive but does not require milker intervention. Since this system is computer driven, keeping a highly accurate and up-to-date database is imperative for it to function properly. Users thinking about this alternative should consider the equipment cost and the labor to place and replace neck bands on cows. They must also consider how cows mistakenly sorted will be handled.

Ventilation

Dairy cows need a constant source of fresh, clean air to achieve their production potential. High moisture levels, manure gases, pathogens, and dust concentrations present in unventilated or poorly ventilated structures create an adverse environment for animals. Stale air also adversely affects milk production and milk quality. Proper ventilation consists of exchanging barn air with fresh outside air uniformly throughout the structures. The required air exchange rate depends on the temperature and moisture level of the outside air, animal population and density.

Most of our modern dairy barns rely on natural ventilation to remove heat and humidity from the animals' environment. Natural ventilation for a barn depends on building openings and building orientation. Barns should be oriented to maximize the airflow into the barn. Since most barns are longer than they are wide, the length of the barn should be positioned perpendicular to the most prevalent wind during the time of the year with the highest heat patterns (i.e. if summer winds normally come from the South, the barn should be built with east-west orientation to take advantage of this wind).

Opening on the sidewall allows air to enter and escape, taking heat and humidity with it. Curtain sidewalls in cold climates close the barn on cold days and control the amount of air movement into the barn. Barns should never be completely enclosed; some barn openings should always be provided. Buildings should be designed with a minimum of one inch of eve opening on each side of the barn and two inches of roof opening at the top for each ten feet of width. During cold weather curtains can be closed to this minimum amount and then opened as temperature increases. During hot weather, the barn should be as open as possible to maximize the amount of air flowing through it. Barns having a 4/12 pitch roof with open ridge and sidewall heights of 12-14 feet are recommended. People considering higher sidewall heights should take into account the tradeoff between additional airflow and the light and heat radiated or reflected into the barn. Wide barns and barns in areas where wind directions change should be designed to be open at both ends. Another ventilation consideration is the width of the barn. Six-row barns are normally wider than 4-row barns which reduces natural ventilation. Chastain (2000) indicated that summer ventilation rates were reduced by 37 % in 6-row bans as compared to 4-row barns.

Another factor to consider when attempting to maximize airflow is the proximity of the barn to other structures or land features. A high hill, wooded area, or another building can serve as a wind shield and prevent airflow to the barn. The higher the obstructing object, the farther it should be from the barn being built. As a general rule freestall barns are built about 100 feet from such obstructions.

Fans and sprinklers

High-producing dairy cows must be kept at comfortable temperatures. Heat stress occurs when cow's heat load is greater than its capacity to lose heat. The heat load includes the cow's own body heat as well as external heat from air movement and temperature, humidity, and solar radiation. Dairy cows do not perspire heavily, so they must rely on evaporation through respiratory heat loss. High respiratory rates result in reduced feed intake and low rumination that negatively affects milk production. Modern freestall barns allow cows to take advantage of shade, fans and sprinklers to reduce heat stress. Figure 1 shows a new freestall barn with fans installed.

Smith *et al.,* 2003, reported that cow cooling by evaporating water off her skin surface is a very effect method of relieving heat stress and decreasing milk lose during times of ever heat. The use of low-pressure sprinkler/soaker and fan systems to effectively wet and dry the cows will increase heat loss from the cow. Dairy cows can be soaked in the holding pen, exit lanes, and on feedlines. The goal should be to maximize the number of wet-dry cycles per hour. In the summer of 2001, a study was conducted at Kansas State University to determine the effects of soak frequency and airflow on respiration rates, skin temperature and vaginal body temperature of heat stressed dairy cattle. Sixteen heat stressed lactating cows (8 primiparous and 8 multiparous) were arranged in a replicated 8x8 Latin Square design. Cattle were housed in freestall dairy barns and milked 2x. During testing, cattle were moved to a tie-stall barn for a 2-hour period from either 1-3 pm or 3-5 pm on 8 different days in late August and early September. Average afternoon temperatures were 88, with a relative humidity of 57 %.

Figure 1. To maximize cow comfort in warm weather, cooling fans should be placed in naturally ventilated Freestall barns.

During the testing period, respiration rates were determined every five minutes by visual evaluation. Skin temperature of three sites was measured with an infrared thermometer and recorded every 5 minutes. Treatments were 4 different soaking frequencies with and without supplemental airflow. Soaking frequencies were control (no soaking), every 5 minutes, every 10 or every 15 minutes. Supplemental airflow was either none or 700 cfm. Each wetting cycle provided similar amounts of water for all treatments. Initial data were collected for three initial 5-minute periods prior to the start of the treatments.

Cows soaked every 5 minutes with supplemental airflow (5+F) responded with the fastest and largest drop in body temperature and respiration rate reducing the initial respiration rate by 47 % at the end of 90 minutes of treatment. Soaking cows every 5 minutes without airflow (5) resulted in a similar response as soaking cows every 10 minutes with airflow (10+F). Soaking cows every 15 minutes with airflow (15+F) and soaking cows every 10 minutes without airflow (10) resulted in similar responses until the last 30 minutes of the study. Supplemental airflow without soaking (0+F) resulted in little improvement over no soaking or airflow (0). Wetting had a greater effect on respiration rate and vaginal body temperature than airflow. However, the combination of wetting and airflow had the greatest effect on the respiration rate and vaginal body temperature. Respiration rates and vaginal body temperature were highly correlated. When cooling heat stressed dairy cattle, the most effective treatment included continuous supplemental airflow and wetting every 5 minutes.

This data suggests that different cooling strategies could be developed for different levels of heat stress. Under severe heat stress soaking every 5 minutes with fan cooling will be the most effective. Under periods of moderate stress soaking every 10 minutes with fan cooling may be adequate. Reducing soaking frequency when temperatures are lower could significantly reduce water usage. Data clearly indicate that the combination of soaking and supplemental fan cooling are superior to either single treatment. If used singularly, soaking cows would have more impact than the use of fans only for cow cooling. These data indicate that about 1/3 of the total reduction in cow respiration rates was due to airflow and the remainder due to soaking. Under periods of severe heat stress, soaking every 15 minutes with airflow is not adequate and soaking frequency must be increased.

Cow cooling with soaking and supplemental airflow is very effective in reducing respiration rate. Many systems may be ineffective because they do not deliver adequate water to soak the cow and/or have an inadequate soaking frequency. To adequately cool cows in a 4-row barn Smith recommends fans be mounted above the cows on the feed line and above the head-to-head freestalls. If 36" fans are used they should be placed no more than 30 ft apart and if 48" fans are used they should be placed no more than 40' apart. Fans should be operated when temperature reaches 70 degrees F, create an air velocity of 4-6 mph and air flow of 800-900 cfm per stall or headlock. Feed line sprinklers should be used in addition to fans. Feedline sprinklers should wet the back of the cow and then shut off to allow the water to evaporate prior to another cycle beginning. Application rate per cycle should be .04 inches/square foot and sprinklers should operate when temperatures exceed 70 degrees F.

High Volume Low Speed (HVLS) Fans

High volume low speed fans are configured as large diameter paddle fans with 10 blades. The blades range from 4-12' long making the diameter of the fan approximately 8-24' in diameter. The fan operates at a speed of between 117 and 50 rpm, respectively. The fans have been used in industrial buildings to circulate ventilation air at a low velocity (3 mph). The have also been used in poultry and livestock barns to provide supplemental cooling of animals by increasing air circulation and air velocity in the barn.

A study conducted in several California freestall barns (Haag, 2001) used HVLS fans placed approximately 60 feet apart, mounted in the middle of the barn over the feed driveway. Research results found no difference in respiratory rated and milk production of the barns with HVLS or high speed fan systems.

Kammel *et al.,* 2003, reported that 20-24' HVLS fans installed in Wisconsin in 2001 were mounted at a height of 16-18', which was typically 1' higher than the overhead garage door at the ends of the center drive-through feel lane, and were approximately 60-70' apart. The cost of these fans was approximately $4,000 to $5,000 per installed fan. Air velocities were measured at a height approximately 6" above the cow's backs when they were lying or standing. Velocities of 200-299 fpm were found over a 20' diameter from the center of the fans which coincided with the feed bunk line. Air velocities of 100-199 fpm were found within 30' of the center of the fan, which coincides with interior freestall platforms. Horizontal velocities of approximately 100 fpm 40' from the fan center, which coincides with the outside alley and freestall platforms. Horizontal velocities in the barn were turbulent similar to a light breeze. Air movement normally was above 100 fpm over most of the barn area. Farmers reported improved air quality, reduced noise, drier alley floors, reduced bird populations, less cow crowding, and they felt the fans reduced loss of milk production during periods of high heat and humidity compared to no fans.

Tunnel ventilation

Tunnel ventilation is a special, yet simple summertime ventilation system. Its goal is to provide air velocity and air exchange concurrently in a barn. Fans called tunnel fans are placed in on gable endwall of a building. Fans are operated to create a negative pressure in the barn, causing air to be drawn into the opposite gable endwall opening. Once in the barn the fresh inlet air travels longitudinally through the structure and is exhausted by the tunnel fans.

For tunnel ventilation to function at it maximum potential, all sidewall, ceiling, and floor openings must be sealed to form the "tunnel".

Tunnel ventilation is not generally an appropriate ventilation system for use in cool and cold periods because it can create cold drafty conditions. Since tunnel ventilation is a summertime only ventilation system, another means of providing air exchange must be in place the remainder of the year. Natural ventilation is the most logical choice. One concern with tunnel ventilation is that air that moves longitudinally through a barn becomes increasingly contaminated with air pollutants and at some point the air may no longer be fresh.

Gooch (2001), states that research has shown that air movement between 400 and 600 fpm can successfully reduce heat stress in dairy cattle. The tunnel fan system for a barn should provide a total fan capacity to achieve this 400-600 fpm air velocity and 1,000 cfm (cubic foot per minute) exchange rate per cow. Inlets should be sized to provide a minimum of one square foot of area for every 400 cfm of fan capacity. Recommended fan controls should turn on a pre-defined band of tunnel fans when the barn air temperature reached 65-68 degrees F and additional fans at 71 to 74 degrees F.

Observation of tunnel ventilated freestall barns shows insufficient air movement may take place in the row of stalls adjacent to a completely closed sidewall. Opening the curtain wall slightly (2-4") by raising the lower curtain from the bottom allows a small amount of air to enter along the length of the barn at cow level.

Since the key to making tunnel ventilation work properly is to move large volumes of air, installing a ceiling in the barn improves the performance of the tunnel ventilation system. This is contrary to the needs of a naturally ventilated barn which uses the high ceiling area to dissipate and discharge stale air. To solve this problem barns that will be tunnel ventilated in hot weather and rely on natural ventilation the remainder of the year can place baffles laterally across the barn at about 100-foot intervals.

Tunnel ventilation systems have measurable capital and operation costs above that of a naturally ventilated system which must be offset by additional milk production in order for the investment to deliver a positive return. Currently naturally ventilated structures that provide adequate air exchange and are outfitted with cooling fans placed over rows of stalls and feeding area remains the preferred system, but tunnel ventilation may be justified in new and existing barns that otherwise would provide poor cow environmental conditions.

Floor surfacing

When building a freestall barn you must pay attention to the concrete surface the cows are exposed to. This surface needs to be smooth so the animals' hoofs are not damaged by rough edges or abrasions, it must have grooves to prevent cows from losing their footing when they slip. After new concrete is poured, all rough surfaces should be removed before cows are exposed to it. Often a large block of concrete or a metal blade is used to remove the abrasive elements on the concrete surface. Figure 2 shows these rough edges being removed with a portable grinder unit.

Concrete grooves are normally placed parallel to the barn's manure alleys to allow water to flow in flush barns and to prevent scraper blades from catching the edges in barns using scraper systems. Grooves should be ½ to 5/8 inches wide and deep, with a sharp edge to catch an animal's hoof when the animal slips. Figure 3 shows how this producer chose to cut groves into the concrete after it was poured and dried. This technique often results in cleaner grooves and may not significantly increase the overall cost of the project. Care should be exercised if grooves are to be floated into the concrete to ensure the resulting grooves have a distinct edge and do not produce abrasive places that can damage an animal's hoof.

Gooch (2003), reported that the cost to bullfloat surfaces with grooves in one direction varies between 10 and 20 cents per square foot. The cost to cut grooves into initially cured concrete is 40 cents per square foot in one direction and 80 cents per square foot for two directional grooving.

Figure 2. The abrasive parts of concrete surfaces should be removed, with a portable grinder or some other means, before animals are allowed to walk on it.

Figure 3. Concrete floors should be grooved to prevent animal slipping.

Here the new concrete floor was being grooved after the concrete has been poured and dried. This technique can insure that each groove has sharp edges that help catch a slipping animals hoof.

Freestall design

Knowledge transfer in cattle husbandry

Cows will use correctly sized freestalls because it is easy for the animal to get up and down and provides a comfortable surface to lie on. To support labor efficiency, stall size should encourage animals to lie straight in the stall with their rump over the back of the stall so manure will fall in the manure alley and not on the stall surface. The sizing of freestalls is determined by the animal's size. Several publications, such as the Midwest Planning Service Dairy Freestall Housing and Equipment handbook (MWPS-7), provide the correct dimensions for each animal size. Selecting a size that accommodates the larger animals in a group in recommended.

The key freestall dimensions to consider are curb height, stall width, stall length, neck-rail height, and freestall divider mounting specifications. If curbs are too low, manure may enter the stall when manure it being removed from the barn and if too high, cows will be reluctant to back out of the stalls. A curb height of 10" is recommended, but normally will be 9.5" if a 2"x10" plank is used to form the curb. Stalls should be wide enough to allow animals to recline and rise easily. If stalls are made too wide, animals will tend to lie at an angle in the stall and may even lie backward in the stall. Both of these situations can lead to dirty cows and additional labor to clean stalls because animals will deposit manure on the stall surface. For the average mature Holstein herd, 45" wide stalls often meet these requirements the best. Larger stalls, 48" wide, may be considered for extremely large or pregnant dry cows. Often 48" stalls are built as a convenience to the builder, whereas, 45" stalls would offer the advantages mentioned, plus allow 6 % more stalls per barn.

Cows prefer to lunge forward when rising, because transferring their weight forward allows them to lift their hindquarters more easily. Eight foot of effective stall length is recommended for mature Holstein cows. Actual stall length can be as little as seven feet if the stall design allows the cow to lunge into an adjacent stall. Lunging forward into an adjacent stall is recommended to encourage animals to lie and rise straight in their stalls. Often a brisket board will be mounted in the front of the freestall to help position the animals when lying down and to provide a bracing point for cows when they rise. Brisket boards are recommended to be installed 66" from the manure curb for mature Holstein cows.

Freestalls should be designed so cows can easily rise. Proper positioning of the neck rail is critical in stall design. The neck rail functions as a guide to position the cow correctly in the stall when she enters the stall or stands in it before or after rising. If the neck rails are too low, cows may be reluctant to enter the stalls, may stand half-in and half-out of the stalls and may have difficulty getting up. If too high, animals may not position themselves correctly in the stall before or after lying down. To identify if stalls are correctly designed, watch cows as they attempt to rise. Normal recommendation is for the neck rail to be mounted at least 42" above the surface on which the cow stands. This dimension must take into account the height of any bedding or bedding mattresses used. Anderson (2002), discussed the characteristics of freestalls which contribute to cow comfort. Based on work in Europe, Canada, and the United States he reports that longer stalls, loops with wider openings, higher neck rail placement, brisket boards no more than 4" above the stall bed, and having the platform in front of the brisket board approximately the same level as the bed are important factors to improve cow comfort and stall usage.

Freestall divider design

There are many different freestall stall divider designs currently being marketed, and they are often referred to by names such as side-lunge, wide loop, straight loop, etc. Whichever stall divider type is selected, its length should allow 14" space between the end of the divider and the manure curb once the stall dividers are mounted (Bickert *et al.*, 2000). Allowing additional space may encourage cows to enter another cow's space, and allowing less space may result in cows hurting themselves as they enter the stall and hit the divider. Remember that if barns have different stall lengths, different length stall dividers should be selected for each.

A critical dimension to consider when selecting a divider is the distance from the top of the stall surface to the bottom of the neck rail which should be 45-50" with an absolute minimum of 42". Recent work by Fulwider & Palmer (2003) has shown that the percentage of time cows lie in a stall increased significantly when the neck rail was raised from 45" to 50" in a mattress based freestall barn (Table 5). If the neck rail is to be mounted on top of the divider and 42" off the stall surface, and the lower rail will be 12" above the stall surface, then the opening should be about 30" (42"-12"=30").

Table 5. Effect of neck rail height on the percentage of freestalls with cows lying in them.

	% Lying Before 1-29 to 2-26-03	% Lying After 4-03 to 5-01-03
Average Stocking Density	96%	94%
Avg Temperature (THI)	21.8F(26.9)	49.1F(49.9)
45" Neck Rail	42.1[b]	43.8[b]
50" Neck Rail	40.0[b]	51.4[a]

[a, b] Percentages with different superscripts differ ($P < .05$).

Another important dimension is the distance from the top of the stall divider's bottom rail to the stall base surface. If the stall divider provides sufficient space for the animal's head, then the bottom rail needs to be high enough to discourage the animal from crawling over it. Producers have reported dissatisfaction with extremely wide loop designs because cows get jammed in them and tend to encourage cows to lie at an angle in their stall. Field observation suggests this bottom rail should be at least 12" above the stall surface. A second consideration with divider design is the amount of space provided at the rear of the stall underneath the divider. Extra space here encourages cows to lie with their butt under the divider and results in cows lying at an angle in the stall. Choosing a divider that is mounted perpendicular to the stall-mounting surface at least 12" past the brisket board is recommended.

Barns with rows of head-to-head stalls allow animals to lung into the stall in front of them. This feature saves space, but can also lead to animals being jammed between neck rails. If this happens a cable or pipe may be required between the rows of stalls. Such a pipe must be high enough to allow the cow to lunge forward, but low enough to prevent her from entering the adjacent stall.

Bedding material choices

Freestalls are often thought of as having two components, a stall base constructed of clay, concrete, wood or some other material and a bedding surface. With deep sand freestalls, sand supports both functions. Whatever components are selected, freestalls should conform to the shape of the cow when she is resting, provide cushion when she is reclining and traction when she is rising. It is recommended that the stall surface be 4% higher in the front than in the back of the stall. This discourages forward movement while resting and improves stall drainage. Many people feel cows actually prefer to lie up hill. One situation to avoid is stalls that are lower in the front than the back. Cows have difficulty getting up under these circumstances. Excessive stall slope may also cause cows to lie incorrectly in their stall.

Many different freestall combinations have been tried over the years with different costs and results. Cows dislike concrete-based stalls unless a thick bedding surface is maintained on top of them. Straw, sawdust, manure solids, and other organic bedding surface materials have been used successfully over concrete bases, but their cost is sometimes prohibitive. Wood-based stalls have not been successful because wood rots and gets slippery when wet. Clay-based stalls can provide cow comfort but require a large maintenance effort since cows dig large holes in the front of the stalls. Producers have used rubber tires for freestall bases. Cows seem to like these tire-based stalls, and bedding requirements are decreased, but getting the tires installed properly is very important. Tires should be of the same size, placed tight together and carefully packed with material to hold them in place. Different types of rubber mats have been tried over the years with mixed results. Some get slippery and promote hock damage, and others have deteriorated in a short period of time.

Mattress-based stalls currently are very popular, and for most producers the choice of freestall bases is between sand and mattresses. Mattress-based stalls normally have some rubber particles, water, or other type of filler that conforms to the animal's body and may offer an insulating effect during cold weather. They have a cover that provides animal traction, may be waterproof, and is durable enough to withstand animal traffic. The initial cost of mattress-based stalls is normally $50-100/stall, and their expected useful life is between 4 and 7 years. Mattress-based stalls need to have some type of absorbent bedding applied to them, but the amount is less than deep-bedded stalls over concrete. The initial investment in sand-based stalls is low, but the labor to fill and maintain them, the cost of the sand used, and the adverse effects the sand has on manure handling and storage results in a high maintenance cost.

Sand versus mattresses - performance and producer satisfaction

A recent survey of Wisconsin producers who increased herd size by at least 40 % from 1994 to 1998 (Table 6) showed no significant difference in DHI milk production or somatic cell counts between those using sand and those using mattresses after their expansion (Palmer and Bewley, 2000). Producers using sand seemed to be more satisfied with cow comfort, and less satisfied with manure management and bedding than those using mattresses. Sand users reported significantly higher satisfaction scores for cow cleanliness and hock damage, whereas mattress users reported significantly higher satisfaction with bedding use and cost and manure management. Culling rates, although not significantly different, showed a slight numeric advantage to sand users. Results of this survey can be found on the University of Wisconsin, Dairy Science Department's web site (www.wisc.edu/dysci/).

Table 6. Average production and satisfaction values of herds using mattresses or sand bedding.

	Freestall Bedding Type	
	Mattresses	Sand
Number Herds	69	145
DHI 1998 RHA Milk(lbs)	22,519	22,539
Avg. Linear SCC	2.88	2.80
Culling Rate (%)	34	32
Cow Cleanliness*	4.12	4.47
Hock Damage*	4.22	4.72
Bedding Use and Cost*	4.25	3.95
	4.32	3.43

* Average satisfaction reported on a scale of 1 (very dissatisfied) to 5 (very satisfied)

An Iowa study, which was designed to evaluate six different freestall surfaces, found that stalls ranked differently by week of trial, with cow preference switching between sand and mattresses (Thoreson, 2000). Sand ranked highest in the summer, but usage declined from summer to winter.

Other research conducted in Europe demonstrated that cows showed definite preference for some types of mattresses and that cow preferences changed over time (Sonck & Daelemans, 1999). It was suggested that cows need time to adjust to some types of mattresses and other mattresses get harder and less comfortable over time.

Table 7 shows the results of a study (Palmer, 2003) which showed that stall base type affects cow preference. This study reported the stall usage for a 4-row freestall barn with 100% stocking rate. Observations of cows lying or occupying stalls (standing or lying) were recorded for a nine month period. Sand and mattress-I (Rubber filled) based stalls consistently had larger stall use percentages; concrete and soft rubber mats consistently the lowest percentages; and mattress-II (Foam filled) and waterbeds percentages were intermediate. The sand based stalls had the highest overall lying percentage, but mattress-I and mattress-II had the highest stall occupied percentages. Cows appear to prefer to stand on soft surfaces provided by mattresses or soft rubber mats to sand stalls or concrete alleys.

Table 7. Cow Preference for different stall base types "Experiment 1" for 4-row barn with 100% stocking.

	Soft Rubber Mat Type I	Waterbed	Mattress-I (Rubber Filled)	Mattress-II (Foam Filled)	Concrete	Sand	Average
% Lying	32.9	45.4	65.2	57.4	22.8	68.7	51.0
% Standing	24.6	7.9	17.0	20.7	8.8	3.3	12.1
% Occupied	64.8	61.6	88.3	84.1	38.7	79.0	70.1
No. Obs.	6727	6727	6727	6727	7688	13454	

The lying percentage advantage of sand over mattress-type-I (68.7 % > 65.2 %) was small compared to the stall occupied advantage of mattress-type-I over sand (88.3 % > 79.0 %). This suggests cows like to lie down on both stall bases, but prefer to spend non-lying time standing in mattress-type-I based stalls rather than on concrete manure alleys. Some stall base types were consistently inferior to others. Lying percentages for concrete and soft rubber mats were always below the average lying percentages. Mattress-I based stalls consistently ranked higher than mattress-II for lying and stall occupied percentages, which indicates not all mattresses are equally desirable to cows and making general statements about "mattresses" may be misleading. The length of time cows are exposed to the different stall bases affects lying and occupied percentages. The waterbed based stalls required a longer adaptation time whereas use of soft rubber mat based stalls in this trial decreased over time.

Table 8 shows the results of a second experiment conducted in the same barn as Experiment 1 (Fullwider & Palmer, 2003). Two different mattress types and three different soft rubber mat types replaced the sand, concrete and waterbed stall bases. Cow preference was strongest for foam and rubber filled mattresses. Cow preference for the two mattress types previously tested and which had been preferred now was intermediate. These two mattress types were installed approximately three years before the other stall bases, so it is not possible to determine if the new mattress types were superior or if the decrease in cow preference of the existing two types was due to an aging effect. Rubber mats were consistently used the least. Differences in stall usage existed between different manufacturers' foam and rubber filled mattresses. Visual inspection shows differences in deterioration and surface levelness of the different products over time. These factors can influence the life expectancy of each product and should be considered along with cow preference when making a buying decision.

Table 8. Cow Preference for different stall base types "Experiment 2" for 4-row barn with 100% stocking.

Stall Base	Exp 2 % Lying 6-19/12-17	Exp 1 % Lying (Ranking)	Exp 2 % Occupied 6-19/12-17	Exp 1 % Lying (Ranking)
Mattress-III (Foam Filled)	62 %^{a}		91 %^{a}	
Mattress Type IV (Rubber Filled)	59 %^{ab}		84 %^{b}	
Mattress Type I* (Rubber Filled)	57 %^{b}	65 % (1)	85 %^{b}	88 % (1)
Mattress Type II* (Foam Filled)	52 %^{c}	57 % (2)	81 %^{b}	84 % (2)
Soft Rubber Mat Type II	51 %^{c}		73 %^{c}	
Soft Rubber Mat Type III	43 %^{d}		64 %^{d}	
Soft Rubber Mat Type IV	42 %^{d}		65 %^{d}	
Average	52 %	50 %	78 %	75 %

[a,b,c,d] Percentages within rows, lying & occupied analyzed separately, different superscripts differ ($P < .05$).

Lighting

Proper lighting is very important because it can provide the proper environment for many tasks, including office work, cleaning milking equipment, treatment of animal, and health of animals. A good working environment is needed to increase working efficiency, comfort, and safety. Amounts of light will vary depending on the tasks that are being preformed. New buildings should be designed to support the lighting needs of the dairy operation.

All fixtures should be watertight and made of corrosion resistant materials. Wiring should be surface mounted cable or non-metallic conduit. When selecting light fixtures initial cost, efficiency, lamp life, color rendition and starting characteristics should be considered. High illumination levels (100 foot-candles) should be planned for milking parlor operator pits, offices, toilets, milk rooms, and animal treatment areas. The average light intensity in barns needs to be at least 15 foot-candles at cows' eye level. Normal recommendation for new facilities is to install lighting designed to deliver 20 foot-candles to allow for the effect of aging and dirt buildup on light futures.

Research trials have shown that supplemental lighting can increase milk production and feed intake. The primary objective of a supplemental lighting system is to provide summertime day lengths year-round. Additional light should be supplied so milk cows are exposed to a constant 16-18 hours of light and a minimum of 6 hours of darkness each day. The expected result of supplemental lighting for commercials herd is an 8 % increase in milk production coupled with a 6% increase in feed intake. Cows do not respond immediately but are expected to adapt in several weeks.

Large herds with several milking groups milked 3x often have difficulty providing six hours of continuous darkness (Smith *et al.,* 2003). Lights may need to remain on at all times to provide lighting for moving cattle to and from the milking parlor. Low intensity red lights may be used in large barns to allow movement of animals without disruption of the 6-hoiur required dark period of other groups.

Dry cows benefit from a different photoperiod than lactating cows. Recent research (Dahl, 2000) showed dry cows exposed to short days (8L:16D) produced more (P<.05) milk in the next lactation than those exposed to long days (16L:8D). Based on the results of these studies, dry cows should be exposed to short days and then exposed to long days post-calving.

Animal grouping strategies

Modern dairy herds are normally large enough to allow the operator to house animals with like needs together and to manage them as groups. These animal groupings can support the nutritional, reproductive or health needs of the animals and the management practices desired. The number of different groups maintained by a dairy is often based on herd size and the milking capacity of the milking system selected. When facilities are being planned, milking cows and dry cows and those with special needs should be designed to support expected numbers of each animal type. Table 9 shows the relative number of each type of animal that could be expected in a 1000 head dairy if a perfectly even calving distribution could be maintained. Since animals are constantly entering and leaving the herd, housing for more than 1000 animal is needed to accommodate heifers waiting to freshen. We must also remember that calving is not always uniform, so the number of individual calving pens shown should be considered a minimum. Housing for the post-calving and early fresh cows should also be designed to comfortably accommodate additional animals if an uneven calving is experienced. Sizing pens for these groups for a 90 % stocking rate is one way to accomplish this.

Table 9. Normal percentages of each type of animal in a dairy herd and expected housing needs of a 1000-cow herd.

Type of Animal	Percent of Cows	Number of Cows
Cows with Saleable Milk:		
↳ Healthy Cows	76 %	760
↳ Slow and Lame	2 %	20
↳ Early Fresh	3 %	30
Total Saleable Milk	81 %	810
Cows with Unsaleable Milk		
↳ Sick Cows	2 %	20
↳ Post-Calving Cows (4 days)	1 %	10
↳ Post-Calving Heifers	0.3 %	3
↳ Maternity (1 day)	0.3 %	3
Total Un-Saleable Milk	3.6 %	36
Dry Cows		
↳ Far-off	12 %	120
↳ Pre-Calving Dry Cows	4 %	40
↳ Pre-Calving Heifers	4 %	40
Total Dry and Close-up	20 %	200
Total Housing Needed	104.6 %	1046

A large modern dairy facility normally is comprised of multiple barns. These barns normally are located far enough apart (approximately 100') so they do not interfere with the natural ventilation of other barns and close enough to minimize walking distance from housing to milking facilities. Normally the herd is considered to be comprised of three types of animals: health milking cows, dry cows, and cows with special needs. Healthy milking cows are normally housed in large barns containing several pens sized to match the parlors milking capacity. Dry cows often are housed in separate facilities away from the milking herd. Cows with special needs are often placed in a barn close to the parlor were they can be monitored and treated (often referred to as a special needs barn).

Herd size, construction costs, and the management approaches planned for the dairy influence the type and location of each barn and what types of animals will be housed in each. Table 9 showed the approximate percentage of each type of animal expected. These values may vary by herd and time of year because of seasonal calving differences, etc. Facilities for large herds should be designed to house each of these groups separately and have enough flexibility to accommodate changes in group sizes. Smaller herds may combine some of these groups.

Special needs facilities

Special needs barns normally house sick and transition cows. A transition cow is defined as a cow in the final stages of pregnancy or the first part of lactation. This transition phase is normally defined to be 2-3 weeks before and 2-3 weeks after calving and is an extremely key period for the cow. The design of this barn should maximize cow comfort and the ability of workers to easily handle animals. Facilities to restrain animals, drover lanes, people passes, proper lighting, lifts, tables for hoof trimming and surgery, supply storage, and calf warming boxes are often incorporated in the design of this barn.

The special needs barn often contains freestall pens, bedded packs, and individual calving pens. The number and size of each pen should be based on the number of animals expected, the owner's management plan, and some additional capacity to accommodate variation in group sizes. Overcrowding of this facility is not recommended.

Figure 4. Cows love to walk on a rubber floor.

Pens with 48" freestalls for pre-calving dry cows, 45-46" freestalls for pre-calving heifers, or bedding packs with at least 100 square foot of pack per animal are recommended. Individual pens (12' x 12' minimum size) and/or group-calving-bedded packs can be used to manage cows before calving. Facilities should be designed to conveniently milk fresh cows and feed calves colostrum. Calving areas should be easily cleaned and sanitized. Gate and access lanes should be provided to easily move fresh cows and calves after calving. Sick cows should be isolated from other animals (by at least 10') to prevent the spread of disease. Freestalls are acceptable for most sick cows, but bedded-packs are recommended for sick that have trouble rising.

Conclusion

To design a dairy facility that enhances management efficiencies and animal welfare is a challenge, but rewarding in that ever effort to increase cow comfort should result in enhanced profits through increased production and/or reduced replacement rates. Proper stall design, floor surfaces, and ventilation are key factors contributing to cow comfort. Proper barn design can enhance ventilation efficiencies and lead to reduced animal stress. Planning ahead to have an adequate water supply to support cow cooling and a manure system that handles the extra water should be primary design considerations. Electrical needs associated with fan requirements should be defined and installed when a new facility is built to reduce installation costs.

References

Anderson, N., 2003.Cozying up to Cow Comfort. Midwest Health Conference, Middleton, WI. Nov 12-13: 11-27.

Bewley, J., R. W. Palmer & D. B. Jackson-Smith, 2001. An Overview of Wisconsin Dairy Farmers Who Modernized Their Operations. Journal of Dairy Science, 84: 717-729.

Bewley, J., R. W. Palmer & D. B. Jackson-Smith, 2001. A Comparison of Free-Stall Barns Used by Modernized Wisconsin Dairies. Journal of Dairy Science. 84: 528-541.

Bickert, W.G., B. Holmes, K. Janni, D. Kammel, R. Stowell & J. Zulovich, 2000. Dairy Freestall Housing and Equipment handbook, Seventh Edition (MWPS-7).

Chastain, J.P., 2000. Designing and managing natural ventilation systems. In: Proc. of the 2000 Dairy Housing and Equipment Systems: Managing and planning for profitability. NRAES publication 129: 147-163.

Dahl, G.E., 2000. Photoperiod management of dairy cows. In: Proc. of the 2000 Dairy Housing and Equipment Systems: Managing and planning for profitability. NRAES publication 129: 131-136.

Fulwidder, W. & R.W.Palmer, 2003. Factors affecting cow preference for stalls with different freestall bases in pens with different stocking rates. Abstracts from the American Dairy Science Association, JDS Volume 86, Supplement 1, Phoenix AR, June 22-26: 158.

Gooch, C.A., 2001. Natural or Tunnel Ventilation of Freestall Structures: What is Right for your Dairy Facility? July. 1-13.

Gooch, C.A., 2003. Floor Considerations for Dairy Cows. Natural Resource, Agriculture, and Engineering Service (NRAS-148), Feb. 1-19.

Haag, E., 2001. Cool Cows for Less. Dairy Today. May.

Kammel, D. W., M.E. Raabe & J.J. Kappleman, 2003. Design of High Volume Low Speed Fan Supplemental Cooling System in Dairy Freestalls. 14pp.

Midwest Plan Service Dairy Freestall Housing and Equipment handbook, Seventh Edition (MWPS-7), 2000. Iowa State University, Ames, IA.

Palmer, R.W., 2003. Cow Preference for Different Freestall Bases in Pens with Different Stocking Rates. Proc. from the Fifth International

Palmer, R. W. & J. Bewley, 2000. The 1999 Wisconsin Dairy Modernization Project – Final Results Report. University of Wisconsin, Madison, WI.

Smith, J.F., Harner, J.P. & M.J. Brouk, 2003. Dairy Facilities-Putting the Pieces Together. Four-State Applied Nutrition and Management Conference. MWPS-4SD16. July 9. 35-45.

Smith, J.F., J.P. Harner, M.J. Brouk, D.V. Armstrong, M.J. Gamroth & M.J. Meyer, 2000. Relocating and Expansion Planning for Dairy Producers. Kansas State University. MF2424, January 2000.

Wagner, A., R. W. Palmer, J. Bewley & D.B. Jackson-Smith, 2001. Producer Satisfaction, Efficiency and Investment Cost Factors of Different Milking Systems. Journal of Dairy Science, 84: 1890-1898.

Health management in large-scale dairy farms

Jos P.T.M. Noordhuizen[1] and Kerstin E. Müller[2]

[1] *Dept. of Farm Animal Health, Faculty of Veterinary Medicine, Utrecht University, P.O. Box 80151, 3508 TD Utrecht, The Netherlands*
[2] *Klinik für Klauentiere, Fachbereich Veterinärmedizin, Freie Universität Berlin, Germany*

Summary

Large-scale dairy farms should be run as a business to be profitable. Often they are being blamed for their bio-industrial features not paying attention to animal health and welfare, and showing high disease prevalence due to their high population density. Usually, various people with different background work on those farms, enforcing management to create internal business units. By applying managerial and organisational science principles each business unit should strive for positive benefits to costs ratios, by setting operational goals, implementing specific monitoring tools, conducting risk and problem analysis, and evaluating performance. These programmes for each business unit should be incorporated at whole farm level. Such an organisation is characterised by a high level of structure, transparency, personnel and process control. At the same time, the farm management should look for added value in the chosen farm advisors like nutritionists, agro-economists and veterinarians. Herd health management in those large-scale operations must be refined up to the level of the business units, following the same principles as named above. As such, the herd health management is focused on processes and functions, where-in risk identification, risk analysis and risk management is more valuable and economically profitable than the curative approach. When one strives after integration of different management components, herd health management can be regarded as a feature of quality management of both the products and the production process. Especially, in this type of farming an integrated quality management approach facilitates not only to achieve a high quality level of production and products but also of environmental, welfare and herd health issues. The different elements as named above are highlighted in the presentation.

Keywords: integrated quality management, process control, risk analyses, critical control points

Introduction

Large-scale dairy operations can be found in many regions of the world, including Europe. They comprise usually more than 500 or 1000 head of cattle, employ many labourers and machines, and exploit a large surface of land. The ultimate goal is to make profit, by reducing costs and or increasing income as visible by the net profits per hectare, per cow and per 100 kg milk.

It is often stated that large-scale dairy operations will hamper animal health and welfare due to the solely production-oriented enterprising, a diminished proportion of time daily allotted to animal observation leading to late clinical disease detection, to rather easy disease transmission between animals and the large groups, and to a difficult management of those different animal groups due to their respective needs. As a result large and sometimes intensive vaccination campaigns are conducted to rule out the most important diseases. In Europe however, there is a tendency to restricted application of such vaccinations.

On the other hand, it can be observed that there are large-scale dairy operations which show high performance in productivity. The latter can only be achieved when for example genetics and nutrition, but surely also when animal health is adequately taken care of. Diseases in cattle and young stock result in production losses (milk yield loss, body weight loss, treatment costs, extra labour, premature culling and possibly death) and health care is therefore an economic issue (Dijkhuizen *et al.,* 1996). Curing diseased animals may easily become more costly than investment in prevention. Health management in large-scale operations should follow the pathway of the business-approach using managerial skills and quality control principles as the basis.

In this paper major features of large-scale dairy operations are highlighted, with special emphasis on organisation and management, as well as the major issues of appropriate management of herd health including hazards identification, risk analysis and risk management. A fictive dairy operation is presented for illustration purposes.

Features of large-scale dairy operations

Large-scale dairy operations can ideally be characterised by their (*entrepreneur like*) approach of applying a set of business principles, a high level of organisation and structure, and their need for added value from farm advisors.

Major principles of such operations are listed in Table 1.

The high level of organisation refers to both the personnel, their tasks and responsibilities, the goal-oriented approach of the operation, the planning of all activities, the evaluation of performance, and the adjustment of practices when ever needed.

Table 1. The ten major features of large-scale dairy operations.

1.	Strong goal orientation
2.	Focus on market position & Demands
3.	Awareness & focus on costs and benefits
4.	Priority setting in management
5.	Analysis & evaluation of performances
6.	Investments are a key issue
7.	Prevention rather than cure
8.	High level of organisation & structure
9.	Operational & Strategic planning
10.	High standards of managerial skills and quality

The organisation is highly structured in different business units in order to be able to manage the farm as a whole. The farm itself shows an optimal lay-out of barns/groups, labour lines, and the handling of inputs and outputs, given geographical conditions and local opportunities and limitations. This is reflected in the general scheme of Figure 1.

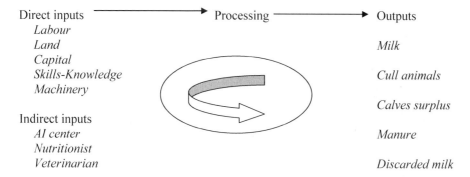

Figure 1. General outline of the inputs, processing and outputs of large-scale farms.

These large-scale dairy operations are usually larger than 500 or 1000 cows and can comprise up to several thousands of cattle. This set-up requires certain conditions with regard to husbandry and (health) management.

Suppose the organisational diagram of that farm looks like the one in Figure 2.

In this fictive situation we can notice that this particular farm has distinguished 5 operational animal units and several supportive units. The operational units are defined on the basis of their specific goals, needs, potential hazards and specific productivity. At the same time one can notice that the husbandry and management of the operational animal units is highly specific in each respective unit.

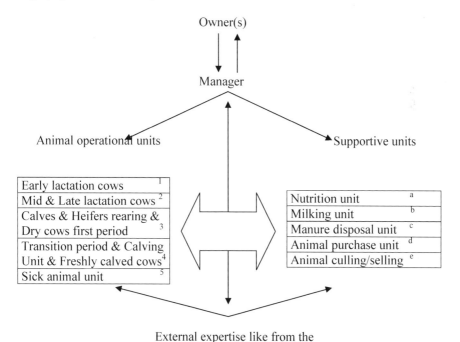

Figure 2. Organisational diagram of a fictive large-scale dairy operation.

The units are managed in a unit-specific way, based on a kind of producer—client relationship.

From a rearing to calving to lactation to dry off and calving perspective, the different animal units are clients of each other: e.g. the calving unit receives cattle from the dry cow unit; they should be able to communicate mutually in order to deliver respectively receive the best "product", the cow.

Product specifications can be defined by the "client", and the provider has to make sure to match these requirements as best possible. Basically all these animal units are also the client of the supportive units and should be approached that way by the latter.

In each unit there is a foreman (a unit coordinator), who has direct managerial contact lines with the manager, as well as with his respective co-workers.

Each business unit has its own targets set, and has described the procedures for achieving these operational goals, while also performance evaluation is part of each unit; personnel responsibilities have defined as well. Mutual relationships between business units have been determined including the way they should interact.

Goal setting, methodology and communication within and between units are among the most important components in this organisation.

Other major components associated with personnel are task descriptions, task assignment, task responsibilities and specific training of individuals, as well as documentation on monitoring animals, farm conditions and on additional animal testing and reporting on achieved performance.

Now the questions can be raised how this could be implemented at the level of a unit and how it could be integrated into one structure, which can be managed and yields high health performance at the same time.

Implementation of managerial principles in a unit: animal health

Let us consider the production process within one particular business unit, e.g. no. 4 the Transition cows, Calving & Fresh cows (see Figure 2).

This unit receives cattle from the Dry cow unit (no. 3), where cows have been dried off and kept for the first few weeks of the dry period. Unit 4 has set requirements for the body condition score of such cows (between 3.5 and 3), the overall health status including the claws and udder (free from infection or disorders; no sour feet; sound quarters), the rumen function status (optimal feed intake as shown by rumen fill scores and faeces quality scores) and an optimal cow comfort before the cows are received from unit 3 (see also Zaaijer & Noordhuizen, 2003). Although both units may agree on the requirements named, it would still be worthwhile to conduct some screening or monitoring of the cows once they enter unit 4. The monitoring or scoring results may prove that unit 3 has delivered the cows in proper state or not, but at the same time are used as starting point for performance evaluation in unit 4. In unit 4 specific targets are set for as long as cows are in this unit.

The production process in unit 4 may look as is presented in Figure 3, where the different areas of attention are being named, as well as the relationships with other areas, like the supportive units.

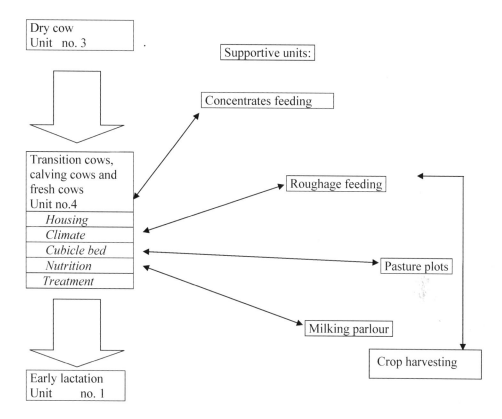

Dry cow
Unit   no. 3

Supportive units:

Concentrates feeding

Transition cows,
calving cows and
fresh cows
Unit no.4

Housing
Climate
Cubicle bed
Nutrition
Treatment

Roughage feeding

Pasture plots

Milking parlour

Crop harvesting

Early lactation
Unit   no. 1

Figure 3. Production process in unit 4.

Within this unit 4 performance targets are set, a monitoring programme designed, a risk identification, risk analysis and risk management protocol defined, treatment protocols described, necessary records installed, and performance evaluation scheduled.

Targets, planning of activities, evaluation and adjustment are 4 components of the managerial cycle.

At the end the cows should be transferred to the early lactation unit (unit no. 1) according to the requirements from that unit. Ultimate objective for this unit 4 is to deliver wholesome milk from healthy cows which show high persistence and peak yield, and longevity against sound profits. Especially this unit 4 has to deal with body condition score, BCS, feed intake variation (rumen fill scores), effects of a negative energy balance around calving possibly affecting subsequent fertility and health status (faeces cores; rectal prebreeding examination; clinical health scoring; see Zaaijer & Noordhuizen, 2003).

Translated into operational goals this means that targets have to be set for BCS (between 3.25 and 3.0 ante partum; between 3.0 and 2.75 postpartum), best feed intake and digestion (no disturbances up to the last day before calving; rumen fill scores around 3; faeces scores of 3 and 2), optimal conditions for rumen adaptation (stress-free housing and handling; no infections; no dystocia) and acceptable disease levels: milk fever (< 2%), clinical mastitis and lameness rates (e.g. <15%), clinical ketosis and acidosis (e.g. <5%), abomasal displacements (<2%). Goals are derived from the hazards in that particular unit.

Hazards in unit no. 4 are e.g. severe negative energy balance and possibly fatty liver with subsequent milk fever, mastitis, lameness, abomasal displacement, ketosis and acidosis, endometritis, anoestrus, repeat breeding.

Each hazard must be coupled to risk conditions contributing to the potential occurrence of that hazard. Some of these risk conditions are general in nature, like hygiene practices and cow comfort, while others are more specific for a particular hazard (e.g. negative energy balance in relation to postpartum ovarian function); some risk conditions can be eliminated or controlled (e.g. poor slatted floor in lameness) while others cannot or difficult (e.g. breed; geographical area; season). Examples of risk conditions for different disorders can be found in Schukken *et al.* (1991), Frankena *et al.* (1992, 1993) and Beaudeau *et al.* (1994).

Monitoring in unit no.4 addresses both cows, risk factors from the cows' environment and management, and available information like milk recording and milk quality data. Monitoring is meant to obtain early signals about possibly deviating performance in a easy, cheap and rapid way and at the same time collect information about possible risk conditions.

➲ Monitoring cows regards the semi-quantitative scoring of body condition, rumen fill, faeces consistency, undigested fibres in the faeces, teat end callosity, claw score, and clinical diseases as indicator of managerial deficiencies, fertility checks (Zaaijer & Noordhuizen, 2003).

➲ Monitoring farm conditions refer to potential risk factors in clusters like housing, nutrition, barn climate, milking, feeding management, contributing to the occurrence of the named hazards.

➲ Available information like milk recording and milk quality checks also provides monitoring sources like somatic cell counts, bacteria counts, antibiotic residues, suspicion of ketosis and acidosis.

All information together should be interpreted in an appropriate way and the synthesis of findings and interpretation could lead to adjustments or interventions on the short and the longer term (Brand *et al.,* 1996).

Treatment of diseased cows should follow the farm advisory list provided by the veterinarian, based on e.g. bacteriological culturing of samples and selecting the antimicrobial products and treatment routes, pharmacologically best suitable for its purpose. Different lists are drawn for e.g. udder health and for calf rearing problems.

In addition to the curing of truly diseased animals, risk assessment and risk management in these large-scale dairy operations should become more and more relevant. Investment in such an approach appears to be economically beneficial.

Integrating the different components at whole farm level: quality control

The farm management has to define the tasks for the different business units, based on the goals and activities illustrated above for unit 4. Task distribution and responsibilities are the next step. The farm personnel should be aware of the fact that they all contribute to the end product of their unit and, hence, to the end products of the farm, and indeed bear responsibility for their part in it. If needed they should receive specific training. They should be motivated to be proud of their performances and checked for that. This is the more participatory type of running the business (*human resource management*). The 3 decision and operating levels comprise the owner and the manager (strategic planning), manager and the foremen (tactical and operational planning), the foremen and their co-workers (operational planning and execution). Each unit should communicate with other units in the provider—client structure, and under an overall management control programme.

Such an approach is highly comparable to the implementation of quality control programmes in other, more industrial branches (Evans & Lindsay, 1996). This is not peculiar because if one considers for example the hazard analysis critical control points, haccp, type of quality control, this also refers to hazards, risk conditions, risk management, performance improvement, quality improvement, performance evaluation, organisation and structure. As was stated in Australia: "haccp is just formalising and structuring what truly good farmers are doing anyway". And since animal health and animal welfare, just like food safety and public health issues, are issues from the production process on a farm, the haccp concept is not only applicable for product and productivity reasons, but equally for animal health and animal welfare, and most probably for environmental issues too (Noordhuizen & Welpelo, 1996; Cullor, 1995 & 1997).

In the haccp-concept too there is a breakdown of the whole production process into different steps (see Figures above), there are hazards identified and the associated risks sampled and possibly managed; documentation and verification are warranted. The haccp-concept regards 7 principles addressing the issues just named (Cullor, 1995). Risk conditions are turned into critical control points, CCP, by meeting all definition criteria presented in Table 2. Because dairy operations, contrary to physical industrial branches, deal with live animals for which standards and tolerance levels are not always easily determined, one would have to define critical management points in addition to critical control points (sometimes referred to as CCP2 and CCP1 respectively).

Table 2. Criteria that jointly define a critical control point, CCP.

Is the site, factor or series located within the production process
Is it associated with the hazard of concern
Does it allow measuring and or observation
Are there a defined standard level and known tolerances
Is it parallelled with corrective measures
If control is lost at this point, corrective measures can restore control
Usually it comprises a cost-benefit assessment of measures and control

All CCPs together determine the on-farm monitoring system; results from monitoring are documented for auditing purposes as well as for operational management.

Examples of critical control points are listed below in Table 3.

The monitoring system should be supported by the documentation of laboratory testing (e.g. bacteriological culturing of mastitic milk; serology; rumen pH; testing results from disease eradication or control programmes) in a planned and structured manner. In that way the haccp –based management programme can be used for demonstrating the herd health status, including the actions taken to maintain or improve that status, to third parties like certification bodies, authorities and consumers.

In addition to health one may add animal welfare and food safety or public health issues to that programme (IDF world dairy summit, 2003).

Table 3. Examples of critical control points on dairy farms.

Milk-born zoonoses
    Purchased cattle should be free from pathogens involved like brucellosis,
        tuberculosis (health certificates). Testing warranted before entry into herd.
        Consequent screening for Staph. aureus mastitis.
        Hygiene barriers for professional people and cars before entry to the farm.
Residues of antimicrobials
    Identification of animals treated with antimicrobials; record keeping.
    Administer appropriate antimicrobials properly (vet prescription)
    Respect withdrawal periods safely.
    Check milk before delivery (bulk tank and or individual cow).
Salmonellosis
    Refuse supply of manure with unknown origin (certificate).
    Hygiene barriers for professional visitors.
    Purchase of cattle with health certificate only.
    Check surface water for fitness as drinking water source.
    Avoid contact of cattle with neighbouring ruminants or wild life.
    Avoid feedstuffs supplied from unknown sources/farms (certificate).
    Strict planning of pasturing cows after manuring.

Conclusions

Large-scale dairy operations are basically not different from family run smaller farms. It is their intensification and scale that require the business-like approach to keep business manageable. Such an approach has all characteristics of how a regular company is managed. They should take advantage of their specific nature, in the sense that they most probably can implement a quality control programme more easily than a small farm. In such a production process oriented quality control programme animal health, animal welfare and food safety issues can be addressed at the same time, e.g. by implementing the principles of the haccp-concept. Hence operational (health) management procedures can be coupled to more strategic quality control issues. Task responsibilities for both can be assigned to different individuals because monitoring activities serve both. Pivotal in this whole play are risk identification, risk management and hence disease prevention (or rather health promotion).

    The fact that there are some large-scale operations showing high performance in terms of e.g. animal health and productivity, while others do not, shows that the differences between these two groups are to be found in the area of overall farm management and organisation, and in the area of controlling and optimising genetics, nutrition and health. This seems in particular valid for situations where massive vaccination campaigns can not be executed, and hence the emphasis must be on disease risk management.

References

Beaudeau, F. et al., 1994. Associations between health disorders during 2 consecutive lactations and culling in dairy cows. Livest.Prod. Sci. 38: 207-216.

Brand, A., Noordhuizen, J.P.T.M. & Y.H. Schukken, 1996. Herd health and production management in dairy practice. Wageningen Pers Publ., Wageningen, The Netherlands.

llor, J. S., 1995. Implementing the HACCP program on your clients' dairies. Vet. Medicine, March issue, 290-295.

Cullor, J.S. 1997. HACCP: is it coming to the dairy? J. Dairy Sci. 80: 3449-3452

Evans, J.R. & W.M. Lindsay, 1996. The management and control of quality. West Publish. Company, St.Paul, Minneapolis, 3rd edition

Dijkhuizen, A.A., R. Morris & R.B.M. Huirne, 1996. Animal Health Economics: rpinciples and applications. The Postgraduate Foundation Publ., Sydney.

Frankena K. *et al.*, 1992. A cross-sectional study into prevalence and risk indicators of digital haemorrhages in female dairy calves. Prev. Vet. Med. 14: 1-12.

Frankena K. *et al.,* 1993. A cross-sectional study into prevalence and risk factors of dermatitis interdigitalis in female dairy calves in The Netherlands. Prev. Vet. Med. 17: 137-144

IDF world dairy summit conference proceedings. 2003. Noordhuizen, J.P.T.M (editor). Quality management at farm level: microbiological contaminants (zoonoses), Bruges, Belgium, September 2003 (in press).

Noordhuizen, J.P.T.M. & H.J. Welpelo, 1996. Sustainable improvement of animal health care by systematic quality risk management. The Vet. Quarterly 18 (4): 121-126.

Schukken, Y.H. *et al.,* 1991. Risk factors for clinical mastitis in herds with a low bulk milk somatic cell count. J. Dairy Sci. 74: 1123-1129.

Zaaijer, D. & J.P.T.M. Noordhuizen, 2003. A novel scoring system for monitoring the relationship between nutritional efficiency and fertility in dairy cows. Irish Vet. Journal 56 (3): 145-151.

Use of longevity data for genetic improvement and management of sustainable dairy cattle in the Netherlands

Rene van der Linde and Gerben de Jong

NRS B.V., Wassenaarweg 30, P.O. Box 454, 6800 AL Arnhem, The Netherlands

Summary

Longevity becomes increasingly important in breeding and management of dairy cattle. In the Netherlands longevity is defined as the time between first calving and last milk recording date of the cow. The average longevity of culled dairy cows in the Netherlands per year of culling varied from 1083 to 1024 days in the past 15 years. No unfavourable systematic trend in longevity can be observed over the past 15 years and results from the last few years even suggest a positive trend. The average lifetime milk production of culled dairy cows increased by 38 percent from 19,593 to 27,109 kilograms from 1988 to 2003.

The average longevity of culled Dutch dairy cows is yearly published in herdbook year reports. A new management information product will be introduced this year in the Netherlands to provide farmers more insight in longevity of their cattle.

Data on longevity is not only used for management purposes but also for improvement of longevity by breeding. In the Netherlands, breeding values for longevity have been computed and published since August 1999 and have also been included in the total merit index. Longevity is weighted 26 percent in the Dutch total merit index, production 58 percent and other health traits 16 percent.

Keywords: longevity, genetic improvement, management information

Introduction

Longevity becomes increasingly important in genetic improvement and management of dairy cattle world-wide. Firstly, longevity is economically important because improvement of longevity will lead to lower costs for rearing replacement heifers. Secondly, longevity becomes increasingly important to consumers because they want healthy cows which stay in the herd for a long time.

In the Netherlands, longevity is defined as the time between first calving and the last test date of a cow and is therefore also called productive life span (PLS).

A new management information product will be introduced this year in the Netherlands to provide dairy farmers more insight in longevity of their cattle. Cattle replacement statistics are published annually in herdbook year reports.

Data on PLS is also used for genetic improvement of longevity. In the Netherlands, estimated breeding values (EBV) for longevity have been published since August 1999 and have also been included in the total merit index.

This paper aims to review the use of longevity data for management and breeding purposes in the Netherlands and their contribution to sustainable dairy cattle.

Average longevity in the Netherlands

In the Netherlands, the average PLS of culled dairy cows is available for the past 15 years. Table 1 shows the numbers of culled dairy cows, average PLS, age at first calving, calving interval, parity and lifetime milk production (LMP) per year at culling.

Table 1. Number of culled cows and the averages for productive life span (PLS), age at first calving (AFC), calving interval, parity and lifetime milk production (LMP) per year at culling of Dutch dairy cows.

Year Unity	Number of cows	PLS days	AFC days	Calving interval days	Parity * 100	LMP Kg
1988	336,918	1110	799	381	325	19,594
1989	348,826	1124	800	382	331	20,472
1990	316,128	1139	798	384	335	21,315
1991	385,465	1137	798	385	334	21,737
1992	360,045	1131	799	387	332	22,106
1993	369,419	1132	800	388	331	22,471
1994	323,257	1151	801	390	334	23,017
1995	369,826	1149	800	391	331	23,359
1996	380,114	1122	801	393	326	23,247
1997	361,624	1110	802	395	322	23,691
1998	369,012	1103	803	397	316	23,842
1999	377,969	1083	803	398	310	23,643
2000	332,618	1092	803	400	311	24,357
2001	336,652	1132	803	401	315	25,265
2002	318,458	1145	803	405	317	25,631
2003	266,805	1204	801	408	328	26,926

PLS increased from 1988 to 1990, was stable between 1990 en 1995 and decreased rapidly from 1995 to 1999. After 1999 PLS increased rapidly from 1083 days in 1999 to 1204 days in 2003. The regression coefficient of average PLS on year at culling is 0.84 days per year ($R^2 = 0.02$), showing a flat trend for average PLS of culled dairy cows over years, despite fluctuations in average PLS of culled dairy cows per year. The fluctuations in average PLS of culled dairy cows per year most likely have to do with the supply of heifers. Between 1996 and 2003 the export of heifers to foreign countries, despite some fluctuations between years, halved on average due to closed borders because of BSE. In this period farmers decided to keep their heifers and to replace older dairy cows with these heifers. Meanwhile the supply of replacement heifers decreased due to more strict environmental legislation. Therefore use of beef bull semen increased strongly in the Netherlands. This resulted in a lower supply of replacement heifers forcing farmers to keep their dairy cows for a longer time, visible in PLS from 2000 onwards. External factors as the export of heifers and environmental legislation can cause fluctuations over years. These external factors have to be taken into account regarding trends in PLS of dairy cattle.

Through the years the average age at first calving stayed very stable between 798 and 803 days. The calving interval gradually increased from 381 to 408 days. The average parity at culling decreased slightly from a maximum of 3.35 in 1990 to a minimum of 3.10 in 1999, but increased to 3.28 in 2003, which is consistent with the changes in PLS. The average PLS of culled dairy cows in 1990 (1139 days) was almost equal to 2002 (1145 days). The average calving interval of culled dairy cows was 384 days in 1990 and 405 days in 2002. The average parity at culling was 3.35 in 1990 and 3.17 in 2002. This illustrates the trend in average PLS of Dutch dairy cows: PLS has a flat trend over years as a result of an increasing calving interval and a decreasing parity at culling.

Although the PLS of culled dairy cows shows some fluctuations from year to year the average LMP per culled cow increases almost every year. The average LMP of culled dairy cows increased by 37 percent from 19,594 kg in 1988 to 26,926 kg in 2003. The regression coefficient of average LMP on year at culling is 388 kilograms per year ($R^2 = 0.94$), indicating a very constant increase in LMP over time. This increase is to certain extent the result of the genetic improvement for milk production in the same period. Despite the strong increase in milk production, both in phenotype and genetics, the regression coefficient of average PLS is even slightly positive. This means that the strong increase in production has had no negative consequences for PLS of dairy cows.

Age at calving versus age at culling

In the Netherlands, average age at calving of dairy cows is available from the milk recording records. As measures of longevity both age at calving and age at culling are used. But these measures can give different results for the longevity of a herd, this is illustrated by Table 2. Suppose that we have three herds with 60 dairy cows and 20 replacement heifers per year and a constant culling policy over time, calving for the first time at the age of 24 months and having a calving interval of 365 days.

Table 2. Average age at calving and age at culling for three example herds with 60 cows each.

Lactation	Herd A	Herd B	Herd C
1	20	20	20
2	20	10	20
3	10	10	20
4	10	10	
5		10	
Average age at calving	38	43	36
Average age at culling	60	60	60

In all the three example herds the average age at culling is 60 months but the average age at calving differs 7 months (43-36). Herd B has the highest age at calving, but not a higher age at culling compared to the other herds. Table 2 shows that the replacement percentage is 33.3 % (20/60) for all the three herds. The longevity measure for these herds should give the same result. Therefore it can be recommended to use average age at culling instead of average age at calving as measure of longevity.

Genetic improvement of longevity

In the Netherlands, EBVs for longevity have been published since August 1999 and have also been included in the total merit index. EBVs for longevity are published four times a year.

Genetic evaluation longevity

PLS is the trait analysed in the genetic evaluation for longevity. The PLS of a dairy cow indicates how long a cow was able to prevent being replaced due to a shortcoming. As cows are kept for the purpose of producing milk, culling based on an unsatisfactory production is called voluntary culling and all other cullings are called involuntary culling. Therefore in the Dutch genetic evaluation for longevity, PLS is adjusted for the within-herd level of production. PLS adjusted for production is also called functional life span. The Dutch EBV for longevity indicates the ability of daughters of a bull to resist involuntary culling, i.e. independent of their level of milk production. Research studies (Van Arendonk, 1985; Vollema, 1998) indicated that selection based on functional life span is more effective than based on realised life span.

The PLS data is analysed using survival analysis (Ducrocq & Sölkner, 1998). With this method PLS is not modelled itself but the hazard of being culled is modelled. Modelling the hazard makes it possible to include also the so-called "censored" records, i.e., records of cows that are still alive at the moment of data collection.

The model used for the genetic evaluation of longevity in the Netherlands includes a baseline Weibull hazard function, time-dependent fixed effects for year-season, parity-stage of lactation, herd size change, intra-herd lactation value (with an economical weighing of kg milk, fat and protein) of the current and the previous lactation, a time-independent fixed effect for age at calving and random effects for herd-year-season, sire, maternal grandsire, genetic groups for maternal granddams and the residual. More details about this model can be found in Vollema *et al.*, (2000). Slight modifications to the model (e.g. inclusion of heterosis in the model) were made afterwards (Van der Linde *et al.*, 2004).

The effects of age at calving, production level and heterosis on days of lifetime estimated in the genetic evaluation are presented in Table 3.

Table 3. The estimated effect of age at calving, production level and heterosis on days of lifetime.

Effect	From … to …	Extra days
Age at calving	24 to 27 months	-44
Production level[1]	100 to 110	256
Heterosis	0 to 100 %	53

[1] Production level is based on lactation value (economical weighing of kg milk, fat and protein), where 100 is the herd average and 110 is 10 % higher.

The difference in life expectancy between cows producing on the herd average or 10 % above herd average (having a lactation value of 110) is estimated on 256 days. The relative production of a cow in the herd is the most important factor related to longevity. The effect of age at calving and heterosis on PLS is moderate.

Direct and indirect information on longevity

The heritability of longevity in the Netherlands is 10 % and therefore it may take some time before enough direct information is available about the longevity of daughters of a bull. Therefore information on correlated (predictive) traits is used to increase the reliability of the EBVs. The published EBV for longevity consists of two components: the EBV based on direct information and the EBV based on predictive traits (indirect information). The amount of information from both sources varies per bull, from 60 percent for very young bulls with no offspring culled to 0 percent for older bulls with thousands of offspring culled. The used predictive traits are the EBVs of rump angle, fore teat placement, udder depth, feet and legs, somatic cell count and interval calving to first insemination.

Use of longevity in the Dutch total merit index

Most countries use total merit indexes to rank their AI-bulls. The total merit index in the Netherlands is called the Durable Performance Sum (DPS). The breeding goal expressed in the DPS is cows with a suitable production, which are healthy and perform a long time on the farm. The DPS ranks bulls on the profitability of their offspring. The DPS is based on the economic values of the most important traits. The current DPS includes production traits, longevity, udder health, female fertility, calving ease, maternal calving ease, viability and maternal viability. Production is weighted 58 %, longevity 26 % and the other health traits together 16 %. Because DPS is an important selection criterion for AI-organisations and dairy farmers in the Netherlands to select bulls, selection on DPS will lead to a genetic improvement for longevity of the Dutch dairy cows.

Genetic trend for longevity for bulls and cows

Selection based on life span gives no response in the realised life span but will lead to less involuntary culling on population level and to an genetic improvement of longevity. The genetic improvement for longevity can be visualised by the genetic trend. Table 4 gives the average EBV for longevity of all Black-and-White Holstein and Dutch Friesian cows per year of birth.

Table 4. The average estimated breeding value (EBV) for longevity of all Black and White Holstein and Dutch Friesian cows per year of birth.

Year of birth	Number	Average EBV	Year of birth	Number	Average EBV
1988	233,159	100.2	1995	248,171	101.1
1989	240,149	100.0	1996	239,656	101.6
1990	234,832	100.0	1997	229,934	101.9
1991	220,091	100.2	1998	232,526	101.8
1992	219,694	100.6	1999	233,952	101.8
1993	226,488	100.7	2000	222,409	102.0
1994	220,928	100.8	2001	194,926	102.4

The EBV for longevity in the Netherlands is expressed with an average of 100 and a standard deviation of 4.5 points. One genetic standard deviation corresponds to 100 days of lifetime, or 50 days at their progeny. The EBVs for longevity for cows are not published in the Netherlands, these are calculated as:

$$100 + 0.5 * (EBV \text{ of sire} - 100) + 0.25 * (EBV \text{ of maternal grandsire} - 100).$$

The unweighted regression coefficient of average EBV per year of birth on year of birth was 0.19 point (2.1 days of lifetime). The decrease in longevity of Dutch dairy cows between

1995 and 1999 can not be observed in the yearly genetic averages. This means that the earlier culling of the cows in these years was due to management.

Use of longevity data in management

Management information products

A new management information product will be introduced this year in the Netherlands. Aim of this product is to make the dairy farmer aware of the replacement policy, longevity and LMP of the dairy cows in the herd and to be able to compare these results with other dairy farms in the Netherlands. This product gives an overview of cattle replacement and LMP of the culled and the present cows per parity and in total in the herd. The total number and percentages of replacement of cattle and the number of heifers that started producing milk will be given. These numbers will be provided on farm level and also as percentile ranking compared to all other Dutch herds. The information product will also contain information on herd averages for production, fertility, udder health and EBVs for production and conformation of all animals in that herd

Cattle replacement statistics

Cattle replacement statistics are published in the yearly Dutch herdbook report. Number of culled animals, their average PLS and herdlife and their average LMP is published. Also the distribution of the average age of living animals per herd is published. In this way Dutch longevity data of dairy cows is made widely available to dairy farmers and consumers.

Conclusions

- ➲ The average productive life span (PLS) of dairy cows per year at culling shows a flat trend.
- ➲ The average lifetime milk production (LMP) of culled dairy cows increased by 37 percent over the past 15 years.
- ➲ The estimated genetic trend for longevity for Dutch dairy cows is about 2.1 days per year.
- ➲ A new management information product gives dairy farmers the opportunity to compare their replacement policy with that of their colleges.
- ➲ Cattle replacement statistics make Dutch longevity data of dairy cows widely available to dairy farmers and consumers.

References

Ducrocq, V. & J. Sölkner, 1998. <<The survival kit 3.0>> A package for large analyses of survival data. Proc. 6th World Congr. Genet. Appl. Livest. Prod. 27: 447-448.

Van Arendonk, J.A.M., 1985. Studies on the replacement policies in dairy cattle. II. Optimum policy and influence of changes in production and prices. Livest. Prod. Sci. 13: 101-121.

Vollema, A.R., 1998. Selection for longevity in dairy cattle. Thesis Wageningen Agricultural University.

Vollema, A.R., Van der Beek, S., Harbers, A.G.F & G. de Jong, 2000. Genetic evaluation for longevity of Dutch dairy bulls. J. Dairy Sci. 83: 2629-2639.

Van der Linde, C., De Jong, G. & A.G.F. Harbers, 2004. Using a piecewise Weibull mixed model in the genetic evaluation for longevity. Interbull Bulletin 32: 157-162.

Labour efficiency and multi-functionality on Irish dairy farms

Bernadette O'Brien[1], Kevin O'Donovan[1 2] and David Gleeson[1]

[1] *Teagasc, Dairy Production Department, Moorepark Research Centre, Fermoy, Co.Cork, Ireland*
[2] *Department of Agribusiness, Extension and Rural Development, Agriculture and Food Building, National University of Ireland, Dublin, Belfield, Dublin 4, Ireland*

Summary

The current inflexibility of dairy systems in terms of labour requirement means that dairy operators cannot easily adopt a multi-functional approach, which may assist in maintaining family farm income. Education in time management is a key element in the promotion of multi-functionality. The purpose of this study was to investigate the labour invested on dairy farms and the feasibility of reducing that labour to provide opportunity for alternative enterprises. Ninety-four dairy farms participated in the study. Proportionally 0.32, 0.28, 0.21 and 0.19 of farms were within milk quota groups 135×10^3 to 250×10^3 litres (Group 1), $>250 \times 10^3$ to 320×10^3 litres (Group 2), $>320 \times 10^3$ to 500×10^3 litres (Group 3) and $>500 \times 10^3$ to $1,500 \times 10^3$ litres (Group 4), respectively. Participant farmers recorded the time taken to perform farm tasks, on consecutive 3 or 5-day periods on one occasion per month. The average dairy labour input per day for farms in milk quota groups 1, 2, 3 and 4 over the 12-month period was 7.0 h, 7.9 h, 9.6 h and 13.3 h, respectively. A daily time saving of 3.0 h and 2.2. h at the milking process and calf care, respectively, was observed on the most efficient compared to the least efficient farms within quota group 1. The data indicated the possibility of reducing dairy labour input on these farms to 3.8 h per day or by 65 %. Well-designed infrastructure and well managed practices employed on farms should facilitate labour efficiency and feasibility of multi-functionality.

Keywords: labour requirements, inflexibility in development, milking, calf care, multi-functionality

Introduction

The contribution of agriculture to national wealth and viability of rural areas is significant. However, it is clear that the sector is in a state of flux. Structural change is continuing within the sector with declining farm numbers and declining employment on farms (Frawley & Phelan, 2002). Currently, the demand for labour on farms is high due to intensive labour-orientated systems of production. In addition, Hennessey *et al.* (2000) have indicated that an expansion of production, of between 60 and 140 % is required if Irish farmers are to maintain incomes. The ability to spread labour costs over a larger milk output can represent a crucial difference between high and low profitability on dairy farms. However, Hennessy *et al.* (2000) also found that labour represented the binding constraint to expansion on 40 % of Irish farms.

The increasing pressure on farm incomes leaves no doubt that the continued existence of many family farms cannot be maintained from farming alone. Part-time farming may offer a mechanism to retain farm families on an increasing number of non-viable farms. Part-time farming could represent a positive impact on the farm household through the contribution of off-farm income to the farm household economy. Part-time farming could also have a positive impact on rural development in that it may retain rural population, alleviate poverty and provide stability in rural areas. The adoption of off-farm work by farm households is an important, well-recognised and growing phenomenon in the EU (Kinsella et al., 2000). The proportion of households with at least one off-farm income source has increased from 36 % in 1995 to 45 % in 2000. Projections indicate that by 2010 some 60 % of farm households will be involved in some form of off-farm work (Frawley & Phelan, 2002). The proportion of farm operators involved in off-farm employment increased from 26 % in 1995 to 32 % in 2000. However, 69 % of these farmers who had off-farm income were engaged in a dry cattle enterprise. But the current inflexibility of dairy systems in terms of labour requirement means that operators within this sector cannot easily take up off-farm employment or develop an alternative enterprise. This must be addressed for the future if a substantial number of dairy farm households are to be maintained.

There is currently a need to establish the patterns of labour utilisation, as well as the influences of facilities, layout and work procedures on labour allocation levels and patterns on farms, in order to elicit constraints and possibilities in relation to labour issues, in a future of potential multi-functionality on dairy enterprises.

Methodology and data sources

The study was conducted with dairy farmers mainly in the Munster region of Ireland, since this area accounted for 65 % of the total manufacturing milk supply in the country. Ninety-four spring-calving dairy farms participated in the study. The farms had spring-calving systems and ranged in milk quota size from 135×10^3 to $1,500 \times 10^3$ litres. Farms were grouped by milk produced into four milk quota groups:
- quota group 1 = 135×10^3 to 250×10^3 litres,
- quota group 2 = >250×10^3 to 320×10^3 litres,
- quota group 3 = >320×10^3 to 500×10^3 litres,
- quota group 4 = >500×10^3 to $1,500 \times 10^3$ litres.

Proportionally 0.32, 0.28, 0.21 and 0.19 of the farms used were within quota groups 1, 2, 3 and 4, respectively. Farms in milk quota groups 1, 2, 3 and 4 had an average milk quota of 212×10^3, 281×10^3, 388×10^3 and 764×10^3 litres, respectively. These farms had an average herd size of 47, 55, 74 and 149 cows, respectively, and an average farm size of 49, 53, 72 and 98 adjusted hectares, respectively.

Data was collected over a 12-month period between February, 2000 and January 2001. All farm operators (farmers and/or farm staff) recorded the duration of the different tasks that they performed throughout the day. Two data recording methods were used. The main method involved a timesheet that was designed to record the total time consumed by 29 different farm tasks for each of 3 consecutive days. The second method involved a Psion organiser, i.e. a hand-held, electronic data logger that incorporated the Observer behavioural package (Noldus Information Technology). On the farms using the Psion, each individual worker recorded data for 5 consecutive days. Sixty-five and twenty-nine farms used data recording methods 1 and 2, respectively.

The 29 farm tasks were incorporated within 10 task categories. This study focused on two task categories, i.e. milking and calf care. 'Milking' was the term used to describe the milking process and incorporated herding cows for milking, milking (clusters on / clusters off), washing up and herding cows after milking. 'Calf care' described the tasks associated with feeding, cleaning and bedding of calves.

'Once-off' survey questionnaires were also completed for each farm participating in the study. The completed questionnaires provided information on facilities and layout and farm practices relating to the milking process, calf care, feeding and cleaning associated with winter housing and waste management on the farm.

Data was processed using the Microsoft Access database management system and analysed using the SAS statistical package (SAS, 1999). Analysis of variance across months was carried out using the GLM procedure. In the analysis carried out, the farm was considered to be the experimental unit from which repeated measures were taken on a monthly basis.

Results

The average dairy labour input per day for farms in milk quota groups 1, 2, 3 and 4 over the 12-month period was 7.0 h, 7.9 h, 9.6 h and 13.3 h, respectively (Table 1). Dairy labour input per day increased with increasing milk quota group (p<0.001).

Table 1. Dairy labour input per day (h) required by dairy task categories on farms of four different milk quota groups.

	Milk quota group					
Group	1	2	3	4		
Av. milk quota (litres)	212×10^3	281×10^3	388×10^3	764×10^3	s.e.m.	Significance
Total dairy labour (h)	7.0^a	7.9^b	9.6^c	13.3^d	0.50	***

[a,b,c,d] means on the same line without a common superscript are significantly different
*** = P<0.001

Labour input on milk quota group 1 farms

It was assumed that milk quota size on dairy farms influences the threat to viability on such farms. Labour input on farms within the smallest milk quota category (milk quota group 1 farms) was examined, specifically, to quantify potential available time for involvement in other enterprises, in order to increase total family farm income. The average total labour input per day for milk quota group 1 farms peaked at 9.6 h in June and gradually declined to 5.4 h in December. When time associated with enterprises other than dairying was excluded, the average labour input per day associated with dairying decreased from 9.0 h in June to 4.4 h in December.

Considerable variation in dairy labour input per day was observed on farms within this quota range. The most efficient 20 % of herds (average quota = 222×10^3 litres) had an average daily labour input of 4.7 h, whereas, the least efficient 20 % of herds (average quota = 228×10^3 litres) had an average daily labour input of 10.1 h (Figure 1). The average production level in terms of daily milk yield per cow of the top 20 % labour efficient herds was 29.0, 25.4 and 21.8 kg/cow/day for the months of May, June and July, respectively. The comparable figures for the lowest 20 % labour efficient herds was 25.9, 23.6 and 21.4 kg/cow/day, respectively. Thus, milk production per cow was not adversely affected by the reduced dairy labour input by the top 20 % labour efficient herds. However, the relatively high dairy labour input of the 20 % least efficient herds may be due to carrying a greater cow number than should be necessary to fill the milk quota. It was observed that the top 20 % labour efficient herds had an average cow number of 42, while the lowest 20 % labour efficient herds had an average cow number of 53 during the month of June. Considerable additional dairy time could be associated with the management of these extra cow numbers at this level of scale.

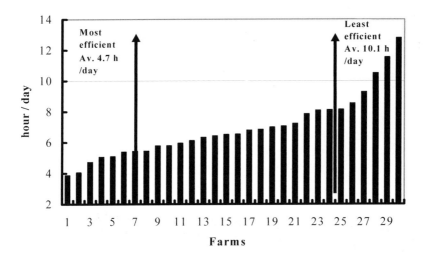

Figure 1. Variation in average dairy labour input per day over 12 months on milk quota group 1 farms.

Benchmarking of milk quota group 1 farms on labour input to specific dairy tasks

Proportionally 0.36, 0.16, 0.12, 0.10, 0.06, 0.11, 0.05, 0.03 and 0.01 of dairy labour time was associated with the task categories of milking process, maintenance (land and buildings), grassland, management, calving and calf care, feeding and checking dairy animals, fertility and miscellaneous over a 12-month period, respectively.

The dairy task requiring the greatest proportion of labour input was the milking process. The average daily labour input for the milking process over a 12-month period, for the 30 farms was plotted in order to establish the month in which labour input was at a maximum. Labour input was highest for the milking process in May. The task of calving and calf care represented a labour demand peak during the month of March. Although calving and calf care accounted for just 6 % of dairy labour input over the 12-month period, calf care alone accounted for 13 % of dairy labour input during the month of March. Thus, the variation in labour input levels in the months of peak labour demand for the tasks of milking and calf care were observed, and potential factors, such as, facilities and practices were compared in order to establish reasons for such variation. The average daily labour input to the dairy tasks of the milking process (May) and calf care (March) for the most efficient (20 %) and least efficient (20 %) herds, together with quota size and cow/calf number for the respective groups are shown in Table 2. (The most and least efficient herd groupings may consist of different herds for the different tasks.)

Knowledge transfer in cattle husbandry

Table 2. Average daily labour input of most efficient (20 %) and least efficient (20 %) herds (within milk quota group 1 farms) to the dairy tasks of the milking process (May) and calf care (March) and quota size and cow/calf number for the respective groups.

	Most efficient 20 % of herds	Least efficient 20 % of herds
Milking process (May)		
Dairy labour input/day (h)	2.2	5.2
Quota (x 10^3 litres)	207	222
Average cow number	40	56
Calf care (March)		
Dairy labour input/day (h)	0.6	2.8
Quota (x 10^3 litres)	209	209
Average calf number	26	30

The milking process

Considerable variation in labour input per day for the milking process during the month of May was observed within this quota range. The most efficient 20 % and the least efficient 20% of herds had an average daily labour input to the milking process of 2.2 h and 5.2 h, respectively (Table 2). In examining the facilities and practices associated with the milking process on the individual farms within the most and least efficient groups, there were a number of factors that could potentially account for the major differences in efficiency. The average number of cows milked per unit was 5 and 8 in the most and least efficient herds, respectively (Figure 2). The most efficient herds had pipeline systems with one operator in the pit. Two of the least efficient parlours had recorder plants and had two operators in the pit during milking, thus, doubling the time associated with milking. There was a greater degree of teat preparation carried out in the least efficient herds. A greater proportion of efficient herds had exit gates operated from any point in the pit. The majority of the most efficient farms had the grazing area in one block, which facilitated cows going to the paddock directly after milking, while the majority of farms in the least efficient group retained cows in the yard until completion of milking, when the cows were subsequently accompanied to the paddock by the drover. There were more instances of mechanized cleaning of yards within the most efficient group, e.g. tractor, pump, slats. The majority of farms in the least efficient group used some degree of hand cleaning which was generally done on a twice daily basis.

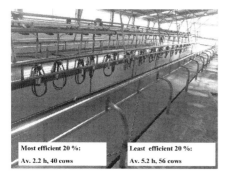

Figure 2. Variation in milking process time (herding+milking+washing) in June.

Calving and calf care

Considerable variation in labour input per day for calf care during the month of March was observed. The most efficient 20 % and least efficient 20 % of herds had an average daily labour input to calf care of 0.6 h and 2.8 h, respectively, in March (Table 2). In examining the facilities and practices associated with calf care on the individual farms within the most and least efficient groups, there were a number of factors which could potentially account for the major differences in efficiency. The majority of farms in the most efficient group transferred milk to the calf house by a trolley type mechanism, whereas, bucket transfer was used on all of the inefficient farms. A minority and majority of farms fed calves individually by bucket on the most and least efficient farms, respectively. The majority of efficient farms cleaned calving houses mechanically and infrequently, while a majority of inefficient farms cleaned calf houses manually using a fork (Figure 3).

Most efficient 20 %:
Av. 0.6 h, 26 calves,
1.4 min/calf

Least efficient 20 %:
Av. 2.8 h, 30 calves,
5.6 min/calf

Figure 3. Variation in calf care time (feeding+bedding+cleaning) in March.

Theoretical dairy labour input

The average for the 20 % of farmers having the highest and lowest dairy labour input per day, for each month of the year was calculated. (Any individual farmer may not have been in the most efficient group for all months or for all tasks within a month.) The theoretical profile of dairy labour input over a 12-month period, incorporating the 20 % least efficient and 20 % most efficient farms (in each month) is shown in Figure 4. The average dairy labour input per day over 12 months for the 20 % of farms with lowest dairy labour input per day and for the 20 % of farms with highest dairy labour input per day was 3.8 h and 10.8 h, respectively. Taking 6 full working days per week, the average dairy labour input for these two scenarios would be 22.8 h and 64.8 h per week or 1,186 h and 3,370 h per year, respectively. Thus, there is potential to reduce labour input by approximately 65 %.

Knowledge transfer in cattle husbandry

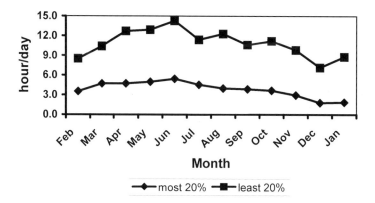

Figure 4. Simulated dairy labour input over 12 months on milk quota group 1 farms – average dairy labour input of 20 % highest and 20 % lowest dairy labour input farms in each month.

Discussion

The increase in labour demand with increasing milk quota group, in this study, was consistent with Adamczuk (1978), who also showed an increase in hours worked with an increase in farm size. Turner and Fogarty (1995) also indicated that an increase in the scale of the farm business on English dairy farms resulted in increased labour demand and annual hours worked per individual. Milking was recorded as the most time consuming task in this study, accounting for over one-third (0.36) of total dairy labour input. This was somewhat higher than the results of Jagtenberg (1999), who recorded that 0.29 of total labour was required by milking. Ordolff (1986) and Sonck (1993) both observed a proportion of 0.30 of labour input required by milking on dairy farms in The Netherlands. Seasonality of production had a large impact on labour requirements (Figure 5). The springtime calving season has traditionally been perceived by Irish farmers to be the period of peak labour input.

Figure 5. Grazing season.

Dairy labour input per day for the quota category investigated (135×10^3 to 250×10^3 l) was 7.0 h. However, a time saving of 3.0 h and 2.2. h per day at the milking process and calf care, respectively, was observed on the most efficient farms compared to the least efficient farms. The data indicated the possibility of reducing dairy labour input on farms to 3.8 h per day (or by 65 % compared to inefficient farms).

Conclusion

In the foreseeable future the option of off-farm employment will most likely be the most effective method for many low-income farmers to supplement farm incomes. In order to accommodate multi-functionality and the operation of the dairy enterprise, labour input on the farm must be minimized without having an adverse effect on productivity or profitability. In this scenario, the number of hours required for satisfactory farm operation and the number of hours available for outside work must be established. The foregoing data gives an indication of minimum time required for farm operations, and mechanisms by which this can be achieved. Well-designed infrastructure and well-managed practices employed on farms should facilitate labour efficiency. Additionally, the use of contractors, casual labour and reserves of family labour for tasks, such as maintenance, calf rearing and winter-feeding of dry cows would further reduce the time commitment on dairy farms and increase the possibility of conducting off-farm employment or engaging in an alternative enterprise.

Education in time management is a key element in the preservation of farming as a core activity through its contribution in improving the quality of life of the operator. Education of dairy farmers as to the realistic possibility for such an operator to become involved in multi-functionality may assist in promoting the incidence of multi-functionality on dairy farms (particularly for lone operators and in instances where it is impractical for the spouse/partner to take part in off-farm employment). It could potentially allow an increase in family farm income while retaining ownership and management of the 'non-viable' farm. Farm incomes have been a major concern in agricultural policy, thus motivating transfers to the farm sector. More recently, off-farm income sources have entered the policy debate as an argument to curtail subsidies (Hill, 1996). Multi-functionality is increasingly seen as a relatively stable adjustment, and thus, there is a policy change towards promoting farm diversification and integrated rural development. Thus, information on the feasibility of pluri-activity /multi-functionality on dairy farms could have a positive impact on policy making.

References

Adamczuk, L., 1978. Research on time allocation amongst the rural population. II. Wies-Wspolczesna 22: 75-84.

Frawley, J. & G. Phelan, 2002. Changing Agriculture: Impact on Rural Development. Proceedings of Teagasc Rural Development Conference, Dublin, Ireland, 20-41.

Hennessey, T., W. Fingleton, J. Frawley, M. Keeney & E. O'Leary, 2000. Changing structure and production potential of Irish dairy farming in the context of quota abolition. Proceedings of Teagasc Agri-Food Economics Conference, Dublin, Ireland, 41-56.

Hill, B. (editor), 1996. Income statistics for the agricultural household sector. Proceedings of the Eurostat Seminar, Luxembourg, 10/11 January 1996.

Jagtenberg, K. 1999. Labour input at low cost farm approaches the objective of 50 hours per week. Praktijkonderzoek-Rundvee, Schapen-en-Paarden 12: 2, 29-31.

Kinsella, J., S. Wilson, F. de Jong, & H. Renting, 2000. Pluriactivity as a livelihood strategy in Irish farm households and its role in rural development. Sociology Ruralis 40: 481-496.

Ordolff, D., 1986. Fully automatic milking: technical aspects and developments. Landtechnik 41: 227-229.

SAS, Statistical Analysis Systems Institute, 1999. ® Proprietary Software Release 8.2 (TM2MO) SAS Institute, Inc., Cary, N.C.

Sonck, B.R., 1993. The modern littered house for dairy cows: too much labour? Labour planning. Labour and Conditions; Computers in agricultural management, Proceedings of the XXV CIOSTA-CIGR V Congress, Wageningen, The Netherlands, 10-13 May: 260-269.

Turner, M.M. & M.W. Fogerty, 1995. Aspects of change in the UK Farm Labour Force. Farm Management 9: (1), 13-24.

Practical experiences with smallholder milk recording in Malawi: a case of Lilongwe milkshed area

Agnes C.M. Msiska[1], Mizeck G.G. Chagunda[1,2], Hardwick Tchale[3], James W. Banda[1] and Clemens B.A. Wollny[4]

[1] Department of Animal Science, Bunda College of Agriculture, University of Malawi, Lilongwe, Malawi
[2] Danish Institute of Agricultural Sciences, Department of Animal Health, Welfare and Nutrition, Foulum, Denmark
[3] Center for Development Research (ZEF), University of Bonn, Bonn, Germany
[4] Institute of Animal Breeding and Genetics, Georg-August University of Göttingen, Germany

Summary

This study aimed at investigating farmer perception on milk recording and factors affecting adoption of milk recording systems. Eighty-six smallholder farmers from six dairy cooperative schemes of Lilongwe were randomly interviewed. A logit regression approach was used to analyze the adoption decision. The results indicated a positive relationship between participation in dairy recording and the individual assigned the recording task (P<0.01), milk recording using simple calibrated containers (P<0.05) and also herd size (P<0.05). A negative relationship was found between recording participation and the following factors: recording using calibrated scales (P<0.01), sale of milk at the informal market (P<0.10) as opposed to the formal market, and use of natural service (P<0.10) as opposed to artificial insemination for breeding. Farmer education level, cattle genotype, and daily milk yield had no significant influence on the adoption of milk recording. Based on the experiences and lessons learnt, a farmer-participatory recording system using simple equipment like calibrated one-litre cups and based on daily recording was designed, successfully tested and adopted.

Keywords: smallholder, milk recording, interviews, participation, socio-economical and physical factors

Introduction

Currently, milk production in Malawi is performed on smallholder and large-scale dairy farms (Figure 1). The major differentiating features of these two dairy sub-sectors are, the holding size, the genotype of cattle raised and the level of management. Recent information (Malawi Government, 1997) indicates that there are about 3600 smallholder farmers who use over 6000 Holstein Friesian x Malawi Zebu cows and about 1700 smallholder farmers who use an unknown number of Malawi Zebu cattle for commercial milk production in the peri-urban setting. In addition to the smallholder farmers, there are 15 private large-scale dairy farms accounting for about 2200 milking cows. The predominant genotype on the large-scale dairy farms is the Holstein Friesian although some of these farms also have few Aryshire and Jersey cattle. The total milk production from both the large scale and the smallholder sub-sectors as at the year 2001 was estimated to be 35 000 metric tonnes per year (FAOSTAT, 2003). Peri-urban smallholder dairy sector supply about 60 % of approximately 9000 metric tonnes of milk a year processed at the three processing plants of Blantyre, Lilongwe and Mzuzu Banda (1996). The smallholder dairy farmers are organised in three milk shed areas around the three major cities of Malawi and operate under corporate approach where at local level farmers belong to milk bulking group. Farmers from within a radius of 8 kilometres km bulk their milk at a cooling centre from where milk processors collect it. Buying of the milk by the processors is in bulk and a bonus is paid for higher bulk quantities. Malawi consumes about 42000 metric tonnes of milk per year (FAOSTAT, 2003).

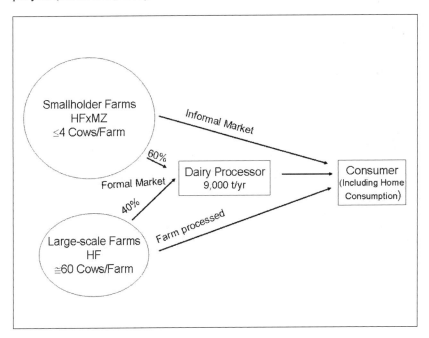

Figure 1. Milk production in Malawi.

With a population of about 11 million people, the estimated average milk consumption is 3.8 kg per capita. This average is very low even when compared with that for Sub-Saharan Africa, which is estimated at 30.8kg per capita (FAOSTAT, 2003). This coupled with the fact that Malawians get 7.46 calories, 1.18 grams, and 0.27 grams per day from milk compared to 52 calories, 2.7 grams, and 2.9 grams of energy, protein and fat for Sub-Saharan Africa the need for improving milk production and the consequent milk consumption in Malawi is heavily pronounced.

Although the smallholder farms play an important role in milk production in Malawi, genetic evaluation studies are currently difficult if not impossible to carry out because of the unavailability of systematically kept records (Chagunda et al.,1998). A smallholder milk-recording scheme that was set in place in 1974 by the Ministry of Agriculture did not pick up. In 1990 the Ministry tried to reactivated the recording system efforts but with no success (Zimba, 1993). To date there is no coordinated animal performance-recording scheme in Malawi. The objectives of this study were to find out why past efforts failed and together with the farmers, based on lessons learnt to develop and test alternative recording systems.

Knowledge transfer in cattle husbandry

Figure 2. Location of Malawi.

Materials and methods

Study area

The study was conducted in Lilongwe milkshed area covering Lilongwe and Kasungu Agriculture Development Divisions (ADD) in the central region of Malawi. There are a total of 18 Milk Bulking Groups (MBGS) in this milk-shed area with approximately 257 smallholder dairy farmers. Farmers from six MBG's were involved in the study. The study was conducted in two phases namely, a survey and a monitoring study. In the survey, a structured questionnaire was administered to a sample 86 randomly selected dairy farmers. The monitoring study involved the development and testing of alternative the recording systems and involved 90 randomly selected farmers. Two multiple trait recording systems were developed and tested in the study area.

These were:

a) a system that required that farmers record on milk yield, information, disease and their treatments,

b) a system where everything in systems (a) plus recording all reproductive and mating activities.

Each system had three recording combinations, (daily, once a week, and once a month), culminating in 6 system–interval combinations. No interventions were introduced in the livestock husbandry practices of the farms. Table 1 presents the outline of the two recording systems. Based on what was learnt in the survey, one-liter calibrated cups were distributed to every participating farmer in the monitoring study. Farmers were advised to measure individual cow milk yield at each time of weigh. Plastic ear tags and names were used for animal identification. Literate farmers were encouraged to do the recording themselves while those who could not read or write were advised to identify one household member that could do the recording.

A two-day training session on the record-keeping format was conducted before the commencement of the monitoring study. Recording sheets were provided in vernacular language. The initial duration of the monitoring study was 4 months. Local extension workers were involved and they assisted in facilitating the exercise.

Figure 3. Countryside of Malawi with cattle.

Data analysis

Descriptive data analysis was followed by logit regression analysis. Data from the monitoring study was analyzed through a Chi-square analysis on the proportion of farmers that did not attempt to record, number of dropouts, and number of those that participated in relationship to the number of farmers at the start of the monitoring period. Correlation of the residuals was used to test the differences in the recording intervals.

Logit regression analysis

The logit regression analysis was performed to test the strength of the relationship between adoption of a recording system and the different variables affecting the decision to record. In the model, the probability of participating was treated as a function of explanatory variables. The explanatory variables that were included in the model were herd size, genotype, insemination system, farmer's education level milk marketing channels, milk yield, milk measuring equipment, and recording task (Figure 4).

The adequacy of the logit model to explain participation was evaluated by chi-square statistic (X^2) while statistical significance of the different variables was tested by t-test. The decision to adopt a technology or not is a binary decision. It is represented as a qualitative variable whose range is actually limited. This variable is limited because it can only take on two values: 1 or 0 (adopt or not adopt) hence the dependent variable in the model of farmers participating in recording was equal to 1 while 0 represented farmers who did not participate in recording.

Knowledge transfer in cattle husbandry

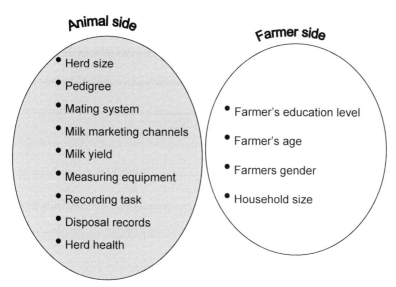

Figure 4. Variables in the hypothesis.

Thus assuming a standard logistic distribution of the error terms, the Logistic model is usually specified as:

$$E(Y_i) = P(Y_i) = e^{\alpha + \beta X}i/1 + e^{\alpha + \beta X}i$$

where:

P_i is the probability of the ith individual with X_i attributes falling in to one of the dependent variable classes,

$E(Y_i) = P(Y_i) = 1$,

$Y_i = 1$ if the individual farm adopts dairy recording and

$Y_i = 0$ if the farm does not adopt recording,

X represents a vector of characteristics or attributes associated with the individual i,

β is the vector of estimated coefficients.

The dependent variable in the empirical model is whether or not performance recording is adopted by the farmer.

Independent variables, besides the conventional socioeconomic ones, include animal related variables.

HERD represents the total number of cattle owned by the farmer and were categorized into three groups of between one and two cows, between three and six, and more than six cows.

GENOT is the genotype of the cows, which could be crossbreed only, Malawi Zebu and Crossbreeds together, and Malawi Zebu.

BRM is the breeding method, which was either artificial insemination (AI), or natural service, both AI and natural service.

EDUC is the education level of the farmer while MMC is the milk marketing channel.

In Malawi, milk is being sold through either through the formal or the informal market.

MYIELD represents the continuous variable of average quantity of milk per day while EQUIP is the milk recording equipment used. Milk in smallholder dairy farms has been known to be measured by a variation of equipment ranging from some as simple recycled bottles, to more conversional equipment like graduated cups or churns and weighting scales.

TASK represents who in the household did the recording. Due to heteroscedasticity (non-constant variance of the error terms) and correlation among the independent variables household size, age, and gender were not included in the final model. On this basis the model estimated in this study was:

$$E(Y_i) = \alpha + \beta_1 HERD + \beta_2 GENOT + \beta_3 BRM + \beta_4 EDUC + \beta_5 MMC + \beta_6 MYIELD + \beta_7 EQUIP + \beta_8 TAST$$

The effect of a change of an explanatory variable with respect to recording performance was predicted using marginal probability, which indicate marginal change in the adoption of a performance recording participation due to a one-unit change in the exogenous variable (Maddala, 1998).

Testing the alternative recording systems

Farmer participation was used to assess the system that was of practical and potential benefit to the smallholder farmers. Three criteria were used, these are:
a) Percentage of farmers that did not attempt to start recording
b) Percentage of dropouts
c) Percentage of farmers that participated until the end of the initial monitoring period

Drop out rate was defined as the proportion of farmers that failed to complete the recording exercise during the course of implementation for reasons other than drying their cows. Farmers who had kept records up to the time their cows had dried off were considered to have participated fully.

Figure 5. Who does the recording?

Results and discussion

Among the farmers interviewed in this study, at least 74.42 % had participated in recording exercises at one point or the other. From the interviews, farmers indicated various reasons why they did not participate in dairy recording (Table 1). These reasons and those reported in literature formed the basis of the logit model testing the factors influencing the recording participation decision.

Table 1. Main reasons for non-participation in recording exercises.

Reasons	n	%
Busy with other activities	17	18.28
Lack of knowledge/Ignorance	15	16.13
Education qualification	12	12.90
Done by non-family members (extension worker / milk buyer)	11	11.83
Low milk yield	10	10.75
No recording materials	9	9.68
Lack of business orientation	8	8.60
Small herd size	7	7.53
No clear objectives/ lack of interest by other stakeholders	4	4.30
Total	93	100.00

The results from the Logistic regression analysis are reported as marginal probabilities of a change in the probability of participation due to a one-unit change in the exogenous variable. Factors with a statistically significant influence on the decision to participate in the recording exercises are; recording task ($P<0.01$), recording equipment ($P<0.01$), herd size ($P<0.05$), type of breeding service provided ($P<0.10$), and type of milk market ($P<0.10$). A further analysis of these factors was conducted to establish to what level the factors influence recording participation.

Factors with a statistically significant influence on the decision to participate in the recording exercises were; recording task ($P<0.01$), recording equipment ($P<0.01$), herd size ($P<0.05$), type of breeding service provided ($P<0.10$), and type of milk market ($P<0.10$). Cow genotype, education level of farmer, and milk quantities produced per day were tested in the model but indicated no significant effect on performance recording. Due to correlation with other factors, household size, age, sex, and participation level of farmer were dropped from the model. A further analysis of the significant factors was conducted to establish to what level the factors influence recording participation and is presented in Table 2.

Table 2. Breakdown of factors affecting milk recording participation decision.

Variables	Regression coefficient	Standard error	T-statistic	Marginal probability
Constant	-4.034	1.326	-3.041***	-
Herdsize				
One or two cows	0.122	0.374	0.325	0.105
Between three and six cows	0.515	0.353	1.459*	0.444
More than six cows	-0.122	0.374	-0.325	-0.105
Recording task				
Self-Recording	1.530	0.643	2.382**	0.776
Spouse-Recording	0.838	0.874	0.959	0.723
Child- Recording	-0.952	0.377	-2.525**	-0.821
Measuring equipment				
Weighing Scale	-0.178	0.558	-0.318	-0.153
Calibrated. Containers	0.466	0.244	1.913*	0.191
Recycled Bottles	0.303	0.252	1.172*	0.162
Mating system				
Artificial Insemination (AI)	0.513	0.230	2.229**	0.212
Both AI & Natural Service	0.105	0.248	0.421	0.032
Natural Service	0.003	0.096	0.027	0.0004
Milk marketing				
Informal Market	-0.121	0.231	-0.525	-0.177

Pearson Goodness-of-Fit Chi Square = 84.624   DF = 69   P = 0.022;
*** $P<0.01$; ** $P<0.05$ and * $P<0.10$.

Recording task

Results indicated that the probability of performance recording participation increased by about 77.6 % when the farmers did the recording themselves as compared to when their spouses did the recording. When spouses did the recording participation decreases by 72.3%. However, if children were assigned to do the recording the probability is reduced further by about 82.1 %. This implies that as farmers shift the responsibility of recording to spouse, children or other household members, the level of recording declines. Expected level in milk recording participation was highest with the farmer doing the recording reflecting the seriousness of the farmer over the recording responsibility. Since the majority of dairy farmers are male, the spouses who might be assigned the recording task are women. In a number of households, women find themselves centrally involved with day-to-day affairs of their homes. Apart from routine household work, there is also labour demand for off-farm activities, social obligations and domestic chores for women (Kapalamula, 1993). The additional activity of milk performance recording might strain labour demand on the spouse, as it would be taken as a supplementary chore in addition to the other household activities. As for children, milk recording, which is mostly done at almost the same time that they are going to school, culminates into a conflict of interest.

Knowledge transfer in cattle husbandry

Milk weighing equipment

A positive and significant level of farmer participation was found when farmers used calibrated containers, P<0.05. Participation in recording increased by about 19.1 % when farmers used calibrated containers compared to when they used other containers. The level of recording increased by about 16.2 % when farmers use recycled bottles. However, the level of participation reduced by 15.3 % when farmers used a weighing scale. Farmers tend to record overall herd milk production levels for cash flow balance at the end of the month. These records are of overall herd production and not individual milk production levels. This is evidence to suggest that farmers realize that there is a need for record keeping although the records are being kept to monitor production and cash output in a farm. While accepting that this is the simplest way of keeping records, overall herd production cannot provide practical evidence relating to individual animal management for individual animal production. While recycled bottles are usually used when selling milk at household level, most farmers take it for granted to document for milk sold to other private markets. Again, use of calibrated containers might be related to knowledge adoption. This means that farmers were not presented with the challenging task of reading calibrations but rather counted the number of cups. While this method has limitations of giving less accurate results, the advantage to smallholder farmers is becoming familiar with the need for record keeping. Findings from this study can be evidenced by Svennersten-Sjaunja et al. (1997) who reported that recording equipment need reliable calibrations systems since systematic errors can influence the result of the measurement drastically.

Herd size

There was a positive and significant relationship between herd size of 3 to 6 cattle and participation in milk recording (P<0.05). For a herd size of one or two cows, the probability of participation in recording was 10.5 %. The herd size from three to six cows had the highest probability of participation in recording of 44.4 % (P<0.005). However, when herd size increased to more than six cows per farmer, the probability of participation in recording dropped by 10.5 %.

The relatively low participation in milk recording in herd sizes of one or two cows can be explained by the fact that most of these farmers are not likely to record as they claim that they are able to keep the transactions in memory. As such these farmers might easily recall the data, while a large herd might require putting down notes.

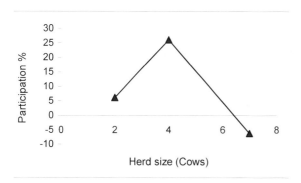

Figure 6. Relation between participation in milk recording and herd size.

Smallholder farmers having one or two cows very rarely keep individual records (Amarasekera, 1998). This is especially true for cattle herds where extreme management system is practiced. As herd size increased from 3 to 6, recording participation also increased. The reason behind might be that as number of animals increased, farmers realized increased milk yield and eventually recording participation also increased. However, recording for individual animals might be a lot of work such that farmers with a large herd (more than 6 animals) might see the exercise as cumbersome, thereby reducing recording participation. In larger herds, good intentions to keep milk production records disappear over a period of time. This is not a positive development. As said by Nicholson *et al.*, 1999, the more animals kept the less likely the intuitive approach to individual animal recording will be accurate.

Mating system

With regard to mating system, results show a positive and significant relationship between the marginal changes in using artificial insemination (AI) and recording participation. By using AI, participation in recording increased by 21.2 %. The rate of participation however only increased by about 3.3 % when farmers used both AI and natural service and increased by only about 0.04 % when farmers used natural service alone. The level of participation when using natural service is a very minimal change that suggests that if use of AI were widely practiced, farmers would be encouraged to keep records than use of natural service. The implication behind this might be that the more farmers use natural service on their cattle, the less likely they are to participate in performance recording. The level of participation in recording for farmers opting for AI might be high because farmers closely observe and handle their animals since the animals are mostly under intensive or semi intensive system of production. As well, farmers using AI are given AI cards, which act as a source of encouragement to the farmer to monitor changes in performance better. In most cases, natural service is widely used for farmers keeping their animals under extensive system of production. In this situation, recording would be a difficult task since the animals are left to graze freely.

Marketing system

The results indicated that farmer participation in recording increases when farmers sell milk through the formal market. A shift from the formal market to a combination of formal and informal market resulted in a decrease of 17.74 % in performance recording. The majority of farmers selling milk on the informal market still maintained some ties with the formal market by supplying some milk to the formal market for membership in the milk-bulking group.

The logit model results also provided information about factors that did not have strong association with adoption of the recording. These factors were: the genotype of dairy cattle being kept by the farmer, education level of farmer, and milk quantities produced per day. Due to alto-correlation with other factors, household size, age, sex, and participation level of farmer were dropped from the model.

Farmer participation levels at different recording intervals

Daily recording of milk production might be an ideal practice in milk recording practices. However, regular interval entries can substitute daily recording (Sudarwati *et al.*, 1995). Results in Table 3 show the level of farmer participation at different recording intervals.

Table 3. Farmer participation levels at different recording intervals.

Farmer Status	Daily recording		Weekly recording		Monthly recording	
	n	% of farmer	n	% of farmer	n	% of farmer
At start	30	100	30	100	30	100
No attempt	13	43.33	9	30.00	15	50.00
Participated	15	50.00	18	60.00	14	46.67
Drop out	2	6.67	3	10.00	1	3.33

n = number of farmers

The greatest participation was observed in the weekly recording interval, (60 %) with 50%, and 46.67 % of the farmers belonging to the daily and monthly recording intervals respectively (Figure 7). Although they were not significantly different, the highest drop out rate was observed in the weekly interval, (10.00 %) followed by the daily recording, (6.67 %) with the monthly interval recording having the least drop out (3.33 %). Results on non-participation also show that about 50 % farmers did not adopt performance recording despite awareness on its importance. On the overall weekly recording seem to have been the most popular of them all.

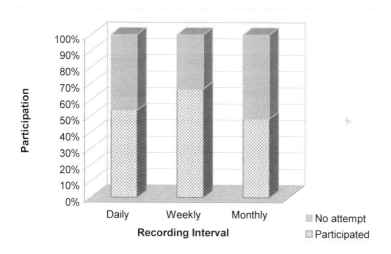

Figure 7. Farmer participation levels at different recording intervals.

Conclusion

The study identifies a number of important socio-economic and biophysical factors that influence the level of participation in milk recording for dairy farmers. A good understanding of these factors is a prerequisite for the introduction and motivation of farmers in milk recording, which should become an integral part of smallholder dairy farming. A simple recording system using easy-to-use equipment, recording all the relevant traits at intervals of not more than one week is feasible and informative on specific farmers' needs.

References

Amarasekera, S.K.R., 1998. Animal Recording in Smallholder Farming Systems - The Sri Lankan Experiences. Ministry of Livestock Development and Infrastructure. In: International Workshop on Animal Recording for Smallholders in Developing Countries. Anand, India. Trivedi, K. R. (eds.), ICAR Tech. Series No. 1: 69-77.

Banda, J. W., 1996. A General Survey of the Dairy and Meat sector in Malawi. Internal Report. Bunda College of Agriculture, Lilongwe, Malawi.

Chagunda, M. G. G., C. Wollny, E. Bruns & L.A. Kamwanja, 1998. Evaluation of the artificial insemination program for small-scale dairy farms in Malawi. Archiv für Tierzucht (Archives Animal Breeding) 41:1/2, 45 -51.

FAO 2003: http://www.fao.org (May 2003)

Kapalamula M., 1993. Comparative Study of Household Economics of Integrated Agriculture-Aquaculture Farming Systems in Zomba District. M.Sc. Thesis, University of Malawi, Zomba, Malawi.

Maddala, G. S., 1998. Limited Dependent and Qualitative Variables in Econometrics. Cambridge University Press: Cambridge, UK.

Malawi Government 1997. Annual Livestock Census. Department of Animal Health and Industry, Ministry of Agriculture and Livestock Development. Lilongwe, Malawi.

Nicholson, C. F., P.K. Thornton, L. Mohammed, R.W. Muinga, D.M. Mwamachi, E.H. Elbasha, S.J. Staal & W. Thorpe, 1999. Smallholder Dairy Technology in Coastal Kenya. An Adoption and Impact Study. ILRI (International Livestock Research Institute), Nairobi, Kenya. 68pp.

Svennersten-Sjaunja, K., L.O. Sjaunja, J. Bertilsson & H. Wiktorsson, 1997. Use of Regular Milk Records Versus Daily Records for Nutrition and Other Kinds of Management. Livestock Production Science 48: 167-174.

Sudarwati, H., T. Djohargarani & M.N.M. Ibrahim, 1995. Alternative Models to Predict Lactation Curves for Dairy Cows. Asian-Australasian Journal of Animal Science 8: 365-368.

Zimba, A. W. C., 1993. A Review of Milk Recording in Malawi. A report for ILCAs Cattle Milk and Meat Thrust. Chitedze Research Station. Lilongwe Malawi.

Online-available milk-recording data for efficient support of farm management

Betka Logar, Peter Podgorsek, Janez Jeretina, Boris Ivanovic and Tomaz Perpar

Agricultural Institute of Slovenia, Animal Science Department, Hacquetova 17, 1001 Ljubljana, Slovenia

Summary

For the purpose of milk recording and herdbook keeping in Slovenia, a central database has been established at the Agricultural Institute of Slovenia. Following the recording scheme, the farmers receive reports of productive and reproductive status of their herds and/or individual cows within a few days. When requested, the reports include urea and somatic cell count info. Milk composition provides valuable information on the management practice of dairy herd. On the basis of fat-protein ratio, animals with digestive disorders and excessive body reserve mobilization, can be detected. Information about urea concentration and its relation to milk protein concentration enables optimization of ruminal nitrogen balance and metabolizable protein supply. Based on the somatic cell count and lactose concentration, farmers are warned about eventual udder infections. Since 2003, the milk-recording data is available online as well. Special application for farmer use was developed. After authorization, the farmers can browse through tabular and graphical presentations of the whole herd or individual animal data. Data on current and previous lactations is available. Reports are in pdf or txt format. Data can be sorted, filtered and exported to other programs for further analysis. The system offers an efficient support to farm management.

Key words: cattle, information system, database, recording, herdbook

Introduction

In Slovenia the history of recording in cattle goes back to the beginning of the twentieth century. In the last decades the organization of expert work in cattle breeding in Slovenia has been based on the organization of Cattle Breeding Service of Slovenia - GSS (Cattle breeding in Slovenia, 1997). Approved breeders' organizations are going to take over the leading role in the year 2005. Milk recording, herdbook service, and genetic evaluation in cattle are the main tasks of GSS. Numerous data is produced by these activities. The methods of collection and processing of data have been changing over time, in connection with the development of information technologies. The objective of this paper is to present the Slovenian Cattle Breeding Information System and Central Cattle Breeding Database with emphasis on online-available milk-recording data used for efficient support of farm management.

Constitution of central cattle breeding database and information system

The establishment of cattle breeding information system and central database goes back to the last decade of the 20th century. Before that, six local and two central databases were held to support GSS activities. Local data bases were maintained by the district cattle breeding service offices. Central databases were run at central institutions of GSS, at Zootechnical Department of Biotechnical Faculty and at Animal Science Department of Agricultural

Institute of Slovenia. In this scheme, joining of the data on the national basis, was quite time consuming. There was an excessive duplication of activities and data storage in the system which led to differences in the data as well. In order to eliminate the duplication of activities and data a single system was established. In the nineties, the development of a new information system was initiated at Animal Science Department of Agricultural Institute of Slovenia. Oracle tools were used for the development of information system and database. Windows operating system was selected for the server to run the relational Oracle database.

Data migration

Building of new database required a migration of data from eight old systems to a new system. The data migration was performed gradually. At first the pedigree data was included in the new system. More than 1.2 million animals were included into the new central database. In 2001, when Slovenia introduced an Identification and Registration (I&R) System (managed by Ministry of Agriculture, Forestry and Food), registration module of cattle information system was already in full functionality. After newborn animal data enters into the cattle database, the I&R data is sent directly into the national I&R database. In addition to the reduced costs, this concept provides important opportunity to reduce the number of errors. At entering the data is checked by a number of business rules based on the information already saved in the database. The next big step in the migration process was relocation of milk recording data. A parallel processing strategy was adopted and herds where switched from the old to the new system on a record-by-record basis. This principle provided the time necessary for resolving eventual discrepancies in the data, before the processing started under the new system. Within a period of two months all herds (>5500) were included in the new system. To conclude, the data migration has been quite time consuming and complex. A great deal of trial-and-error was required before all inconsistencies were resolved. The new information system is fully operational and incorporates national cattle herdbook and performance data.

Structure of the cattle breeding information system

The current overall design of the system is schematically presented in Figure 1. As a tool of communication with the central cattle database, two types of applications were developed: classical client-server application and several web applications. The client-server application using internet consists of modules that enable users to have select, insert and update access to the data base. Internet applications are accessible through web browsers. They enable data input, alteration and a wide range of different reviews and analysis. After login the user gets access to the portal procedures. The menu is dynamic and can be personalized depending on the permission one user has. To be more flexible in supplying information to different users, specific modules were prepared. Numerous management reports for the use by service organizations are available. These reports are created either on demand or according to an agreed schedule and can be prepared in MS Excel, HTML or PDF format. Beside the internet applications for the use on personal computers, special applications for the use on the mobile phone (WAP) or pocket PCs were developed. Since July 2004 the data can be reviewed and inserted into the database using mobile phones or PDA directly in the barns.

The users are divided into two groups depending on their access rights. In the first group there are users from central GSS institutions, district GSS institutions (district offices of Chamber of Agriculture and Forestry of Slovenia (KGZS), and insemination centers. Breeders, agricultural advisory service staff, governmental officials, and other organizations form the second group of users. The first group of users is allowed to use all the types of applications. Depending on their access rights the users from this group can perform all the actions on data. The users from the second group use those internet applications that enable selection and review of data only.

The core of the information system is the Oracle data base, which is located at the Agricultural Institute of Slovenia. It is now being used for information needs of milk and meat recording, type classification, herdbooks, breeding bulls, bull dams, reproduction, and it is a source of all pedigree and most performance data for national cattle genetic evaluation in Slovenia. From this database the data for various publications (catalogues), reports and research and the data for different national needs is prepared.

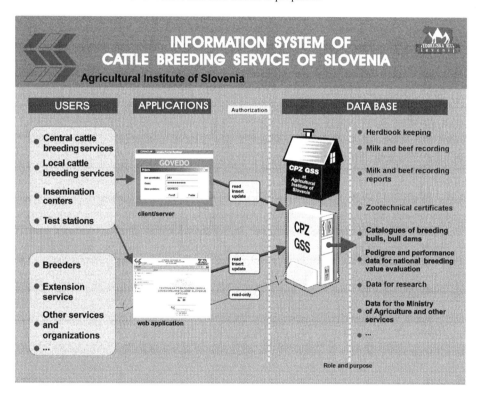

Figure 1. Cattle information system.

Portal for breeders - a new tool for efficient support of farm management

At the time when the new data base and client-server application were set up, the use of internet became more and more widespread. When developing the internet applications, great importance was laid on modules for farmer's needs. This part of web application has been named 'portal for breeders'. Every breeder included in a recording scheme, as well as a breeder included only in I&R, have access to their own data through this system. After the authorization, the farmers can browse through tabular and graphical presentations of the whole herd data or individual animal data. The data is available for all animals that are (or were) in a particular herd. Some of the most important reports present pedigree data and milk recording data. From the tabular views there are further links to the production, reproduction and other relevant data. Current and previous recording data, as well as data on the running and already finished lactations may be reached on portal for browsing. The data is represented with different colors and therefore most important parameters can be easily accessed. Also a report for the review of expected calvings based on insemination data is available. For any herd the breed fraction information for active animals is available. Information about the sires is available in a form of PDF catalogue of a breeding bull. A module for a mating plan, which takes into account the potential inbreeding is under development.

There are three different ways to deliver the recording report. All breeders in the milk recording receive the reports within a few days after recording by post. Optionally, the same reports could be received by E-mail or downloaded from 'portal for breeders'. In such case, the farmers in milk recording scheme can use the recording results immediately after the data processing is finished. When requested, the recording reports also include urea and somatic cell count info. Both reports are sent by mail and could also be accessed via 'portal for breeders'. Milk composition provides valuable information on the dairy management practice of the dairy herd. The somatic cell count info is a good sign of potential subclinical mastitis. The urea concentration and its relation to milk protein content enable the optimization of the ruminal nitrogen balance (three levels: poor, good, excessive) and the metabolizable protein supply (three levels: poor, good, excessive). Cows are graphically arranged into nine groups according to their milk urea and protein content. For every group there are brief guidelines showing how to improve the diet. These can be obtained simply by clicking on the group. This analysis is represented as a graphical report depicting each cow individually. Another graphical report shows digestive disorders and excessive body reserve mobilization and is based on fat-protein ratio. Cows are arranged into nine groups showing whether metabolizable protein/energy supply is poor, good or excessive on one hand and whether bulkiness of diet is poor, good, excessive on the other hand. The brief guidelines for diet improvement are here at hand as well. A more detailed manual on dairy cows feeding has also been published (Babnik *et al.*, 2004).

Conclusion

The new cattle information system greatly facilitates the provision of breeding and management information to cattle breeders. The benefit of this system is a better und regular information service for the breeders, technicians and other services. Fast and reliable information is crucial for good dairy farm management and the 'portal for breeders' is becoming an essential tool to access this information. The increasing traffic on the site and the feedback from the users proves its necessity. Try out the link https://gss.kis.si, on which a demo access is possible.

References

Cattle breeding in Slovenia, 1997. Cattle Breeding Service of Slovenia: Ministry of Agriculture, Forestry and Food, Ljubljana, Slovenia.
Babnik, D., J. Verbic, P. Podgorsek, J. Jeretina, T. Perpar, B. Logar, M. Sadar & B. Ivanovič, 2004. Priročnik za vodenje prehrane krav molznic ob pomoči rezultatov mlečne kontrole, (Raziskave in študije, 79). Ljubljana: Kmetijski inštitut Slovenije, Ljubljana, Slovenia.

Influence of work routine elements of milking on milking parlour performance

Bernadette O'Brien[1], Kevin O'Donovan[1,2] and David Gleeson[1]

[1] *Teagasc, Dairy Production Department, Moorepark Research Centre, Fermoy, Co.Cork, Ireland*
[2] *Department of Agribusiness, Extension and Rural Development, Agriculture and Food Building, National University of Ireland, Dublin, Belfield, Dublin 4, Ireland.*

Summary

This study measured work routine times (WRTs) for different work routine elements of milking and established the factors affecting milking parlour performance. Measurements were conducted in a research milking facility. The time taken to carry out each element of the milking routine was recorded for 50 rows of cows (5 rows x 10 milkings) over a five-day period. Measurements were conducted in a 14-unit parallel, mid-level parlour, with swing-over arms, automatic feeding and sequential bailing. Measurements were taken with one operator when average milk yield/cow was 27.4 litres/day. A personal organiser was used for data recording. The maximum predicted number of cows milked/operator-hour was 79 and 88 with a routine incorporating a complete teat preparation technique (washing, pre-milking, drying of teats) and a 3 s time interval/cow for miscellaneous tasks, in the absence and presence of automatic cluster removers (ACRs), respectively. Comparable numbers for a 5 s time interval/cow for miscellaneous tasks were 76 and 85, respectively. Predicted milking performance was increased to 87, 96, 110 and 116 cows/operator-hour when the pre-milking (drawing foremilk); drying; pre-milking, washing and drying; and washing, drying and dry wiping elements were eliminated, respectively, in the presence of ACRs. Predicted optimum unit number for these routines was 15, 16, 19 and 20, respectively. The ability to accurately predict parlour performance associated with changes in milking parlour size/design and milking routine will be necessary for future herd-size expansion.

Keywords: milking, parlour performance, parlour design, work routines, expansion

Introduction

Milking performance is a most important determinant within the dairy farming system and will become an issue of increasing concern, particularly in the light of increasing herd size. In addition, milking performance must be addressed in the context of good milk quality and cost-effective production. The overall objective is to harvest the maximum volume of milk with the least amount of labour, under the least stressful conditions for the person and cow. The number of cows milked per operator-hour is the best measure of performance of both the operator and the milking facility. A focus on milking performance requires consideration of the following issues.

Current labour input associated with milking

The Moorepark labour study has highlighted the very high labour requirement associated with milking. The complete milking process defined as herding time+milking time [in parlour]+washing time accounted for approximately 33 % of dairy labour input over a 12 month period (O'Brien *et al*., 2003). Thus, improvements in milking performance may have a greater influence than any other aspect of the dairy farmers work, on overall farm labour requirements. So it is appropriate to investigate existing obstacles and limitations to efficient milking and to investigate the role of technology in reducing the time associated with milking.

Factors affecting milking performance

Milking performance is influenced by factors including adequate milking units with minimum vacuum losses, an efficient work routine time (WRT), fast cow flow at entry and exit, a reliable drafting system and stall work that gives good cow control. It is extremely important that the operator does not have to leave the pit during milking time. Upgrading of many parlours in respect to these characteristics is required.

Work routine time (WRT)

There are three main factors that determine the performance of a milking installation namely; individual cow milking time, the number and arrangement of the milking units and the operator's work routine. WRT represents the average time taken by the operator to complete the series of routines associated with each cow. The operator's WRT depends mainly on the number and duration of the elements in the routine and on the degree of automation (Armstrong & Quick, 1986). A typical work routine includes the following tasks; cow entry, teat preparation, cluster attachment, checking of cow for completion of milking, detachment of cluster, teat disinfection, cow exit and miscellaneous (fixing cup slips, rest, etc). In order to achieve optimum milking performance, the number of milking units, the milking time per cow and the milking work routine have to be co-ordinated. A greater number of milking units will allow potential for a greater number of cows to be milked per hour if the operator has idle time during milking (Hansen, 1999). Reducing WRT would only increase idle time unless sufficient milking units are available. Alternatively, too high a number of milking units would lead to increased milking unit idle time, and consequently either over-milking (in the absence of automatic cluster removers [ACRs]) or idle milking units (in the presence of ACRs), unless WRT is shortened (Whipp, 1992). Both parlour performance and milker performance must be optimized in order to achieve maximum milking performance (Klindworth, 2000).

As herd-sizes continue to increase and the dependence on hired labour is higher, a greater focus on working conditions and ergonomics will also develop. Current information on detailed parlour WRTs, and thus, limitations to optimum performance is scarce. Some anecdotal information exists, but this does not provide dairy operators with information that would allow them to predict accurately, the parlour performance resulting from changes in milking parlour size and design, and changes in milking procedures and routines. The objective of the current study was to measure WRTs for the different work routine elements of milking and thus, establish the main factors affecting milking performance in modern Irish milking parlours.

Materials and methods

Work routine times were recorded for different milking activities in a research milking facility. Work routine measurements were carried out on a 70 cow herd. The herd was milked by one operator in a 14-unit, parallel, mid-level parlour, with swing-over arms, automatic feeding and sequential bailing. ACRs were in place and these could be switched on or off as required. WRT measurements were taken for a full range of milking activities over the May / June period (2002) when cows were yielding an average of 27.4 litres per animal per day. The measures recorded were:

a) cow entry,
b) cow exit,
c) washing teats,
d) drying teats,
e) washing and drying teats as one task,
f) drawing foremilk,
g) dry wiping teats,
h) attaching clusters,
i) detaching clusters,
j) changing clusters,
k) disinfecting teats and
l) washing cow standings.

Each measurement incorporated the time taken to complete the task for a full row of cows. This time was divided by the number of cows in the row when calculating work routine times per cow. Measurements were taken for 50 rows in total (5 rows for each of 10 milkings). Recordings were taken at subsequent morning and evening milkings for 5 days. Cows were at pasture during the period of measurement. A hand-held data logger was used for data recording by the continuous timing method (Armstrong & Quick, 1986). The duration of each individual activity was calculated by subtracting its starting time from its ending time.

Results

The time associated with different elements of the milking work routine together with the predicted number of cows that could be milked per operator-hour with a range of different work routines and the optimum number of milking units for different milking routines (based on unit times) is shown in Table 1. A unit time of 10 min was used (Klindworth, 2000) which did not allow for idle unit time or over-milking. Consequently unit time was assumed to be equal to milk-out time.

As various elements of the work routine were automated or excluded, WRTs decreased while the number of cows milked per operator-hour increased. A time interval of 3 s or 5 s per cow was allowed for miscellaneous tasks (fixing cup slips, rest, etc.) in each of the work routines. Milking routine A involved cow entry, washing teats and drawing foremilk as a combined task, drying teats, changing clusters, disinfecting teats, washing cow standings, cow exit and miscellaneous (5 s). Milking routine A allowed for a maximum predicted milking performance of 76 cows per operator-hour to be milked. Milking routine B assumed that ACRs were in place, thus eliminating cluster removal and including cluster attachment. Milking routine B allowed a maximum predicted milking performance of 85 cows per operator-hour. Predicted milking performance per operator hour increased to 87 cows per operator-hour when the task of drawing foremilk was excluded from the routine (milking routine C). Predicted milking performance increased to 96 cows per operator-hour when the routine included washing of teats and drawing of foremilk but excluded teat drying (milking routine D). In milking routine E all teat preparation tasks were excluded except for dry wiping of teats. This allowed a predicted milking performance of 110 cows per operator-hour to be achieved. When all teat preparation practices except drawing of foremilk were excluded (milking routine F), the predicted number of cows milked per operator-hour increased to 116.

Predicted optimum unit number for milking routines A, B, C, D, E and F was 13, 15, 15, 16, 19 and 20, respectively.

Table 1. Time associated with different elements of milking work routines together with the predicted number of cows that could be milked per operator-hour with a range of different work routines and the optimum number of milking units for different milking routines (based on unit times).

Milking routine	A	B	C	D	E	F
Cow entry (s/cow)	3.4	3.4	3.4	3.4	3.4	3.4
Washing teats + drawing foremilk (s/cow)	11.5	11.5	-	11.5	-	-
Washing teats (s/cow)	-	-	10.0	-	-	-
Drawing foremilk (s/cow)	-	-	-	-	-	5.1
Dry wiping teats (s/cow)	-	-	-	-	6.5	-
Drying teats (s/cow)	5.0	5.0	5.0	-	-	-
Attaching clusters (s/cow)	-	10.1	10.1	10.1	10.1	10.1
Changing clusters (s/cow)	14.8	-	-	-	-	-
Disinfecting teats (s/cow)	1.9	1.9	1.9	1.9	1.9	1.9
Cow exit (s/cow)	1.9	1.9	1.9	1.9	1.9	1.9
Washing cow standings (s/cow)	3.9	3.9	3.9	3.9	3.9	3.9
Miscellaneous (s/cow)	5.0	5.0	5.0	5.0	5.0	5.0
WRT (s)	47.4	42.7	41.2	37.7	32.7	31.3
WRT (min)	0.79	0.71	0.69	0.63	0.55	0.52
Max predicted cows/h	76	85	87	96	110	116
Unit time (min)	10.0	10.0	10.0	10.0	10.0	10.0
Optimum number of units	13	15	15	16	19	20

Discussion

Cow entry and exit times are becoming increasingly important as parlour length increases due to increased unit number. Efficient cow flow in the research farm milking parlour in this study resulted in it being unnecessary for the operator to leave the pit during milking. This may be due to the presence of the indoor feeding system in the parlour. Good cow entry and exit times of 3.4 and 1.9 s/cow respectively, were probably due to wide, bright funnel-shaped entrances, which were free of any obstacles and wide, straight exits. However, unlike many parlours, no overlapping of entry and exit was possible due to the bailing system in place.

The time associated with teat preparation varied considerably, depending on the methods used. Milking parlour performance or output is very much influenced by cow preparation practices. Milking routines with full cow preparation and minimal cow preparation had predicted milking performances of 85 and 116 cows/h, respectively. Meanwhile Smith *et al.* (1998) indicated that, minimal pre-milking teat preparation (pre-dip) and cluster attachment took 14 s/cow compared to 25 s/cow for full pre-milking teat preparation (pre-dip or spray, strip and wipe) and cluster attachment. That study also showed that modifying the milking procedure from cluster attachment (9 s/cow) to a full pre-milking teat preparation and cluster attachment (25 s/cow) reduced predicted milking performance from 102 to 56 cows/operator-hour. Armstrong *et al.* (1994) found that the use of a wash-pen increased cow throughput by 8-20 % by reducing preparation times. Minimal cow preparation may not have as significant effect on TBC and milk sediment levels when cows are at pasture compared to indoors. However, from consumer perception and health and safety viewpoints, the issue of preparation of cows for milking has to be addressed. Automation of preparation procedures would speed up milking and cow throughput significantly. Automated teat disinfection is not common in Irish parlours with most operators using hand-held teat dips or sprays or drop-down sprayers. Fox and Smith (1986) found that teat disinfection took 2 s/cow in rotary parlours and between 2 and 3 s/cow in herringbone parlours. Teat disinfection in this study took on average 1.9 s/cow. Ginajlo (1985) stated that spraying was faster than dipping, however the spraying system if installed would require additional capital investment.

In this study, attaching clusters and changing clusters (no ACRs) took 10.1 and 14.8 s/cow, respectively. Therefore the introduction of ACRs reduced work routine time by 4.7 s/cow. While the time taken to remove clusters manually may not be significant, the time spent by operators in making decisions as to when to remove clusters is significant (Klein & Hakim, 1994). Other benefits of ACRs include potential reduction in instances of over-milking and a reduction in the risk of sediment problems in the milk (due to cluster fall-off) (Klindworth, 2000).

Figure 1. Large Holstein-Friesian herd on pasture in Ireland needs to be milked in an efficient way too.

The particular requirements of the individual dairying enterprise and the opportunity cost of labour must dictate the level of automation decided on. If a high level of automation is installed, then it must be ensured that it is reliable and dependable and can be operated by a person of reasonable skill. The cost of automation will depend on the degree of automation.

As herd-sizes are expected to increase in the near future, redesign or construction of new parlours will be necessary. The choice of milking parlour should be directly related to the number of cows being milked currently as well as the herd-size envisaged for the future. Larger herd sizes will lead to a greater focus on time, working conditions and ergonomics associated with milking (Figure 1). In an environment where labour is scarce, limitations due to both time input and ergonomics must be minimized. Thus, it is important that maximum potential milking performance be achieved from new milking installations and from changes in existing milking parlour size and design. Herd owners should focus on a unit number adequate for current efficient milking, while allowing sufficient scope for future expansion and automation.

References

Armstrong, D.V. & A.J. Quick, 1986. Time and motion to measure milking parlour performance. Journal of Dairy Science 69: 1169-1177.

Armstrong, D.V., J.F. Smith & M.J. Gamroth, 1994. Milking parlour performance in the United States. Proceedings of the 3rd International Dairy Housing Conference, Orlando, 2-5 February 1994: 59 – 69.

Fox, J. & P. Smith, 1986. More efficient milking. Proceedings of the 38th Ruakura Farmers Conference, 1 –3.

Ginajlo, M., 1985. Time study of milking and cleaning routines in the Maffra district during spring. Proceedings of the Dairy Production Conference, 1985. Australian Society of Animal Production. (Ed. T.I. Phillips): 529 – 32.

Hansen, M.N., 1999. Optimal number of clusters per milker. Journal of Agricultural Engineering Research 72: 341 – 346.

Klein, J. & G. Hakim, 1994. "A review of labour productivity in milk harvesting in Australia". National Milk Harvesting Centre, Agriculture Victoria, Ellinbank, Victoria, Australia.

Klindworth, D., 2000. Working Smarter Not Harder – benchmarking your labour use in the milk harvesting process. Agriculture Victoria Ellinbank, RMB 2469, Hazeldean Road, Ellinbank, Victoria, 3821, Australia

O'Brien, B., K. O'Donovan, J. Kinsella, D. Ruane & D. Gleeson, 2003. Factors affecting labour efficiency of milking. Book of abstracts of the 54th Annual Meeting of the European Association for Animal Production: 284

Smith, J.F., D.V. Armstrong, M.J. Gamroth & J. Harner, 1998. Factors affecting milking parlour efficiency and operator walking distance. Applied Engineering in Agriculture 14: 643 – 647.

Whipp, J.I., 1992. Design and performance of milking parlours. In: "Machine Milking and Lactation". Bramley A.J., F.H. Dodd, G.A. Mein & J.A. Bramley (editors), Newbury, Bershire, UK, Insight Books: 285-310.

Transfer of knowledge to practice in Slovenia

Marija Klopcic[1], Joze Osterc[1], Marko Cepon[1] and Branko Ravnik[2]

[1] *Biotechnical Faculty, Zootechnical Department, Groblje 3, 1230 Domzale, Slovenia*
[2] *Chamber of Agriculture and Forestry, Mikloŝiĉeva 4, 1000 Ljubljana, Slovenia*

Summary

The extension service and experts from the area of milk recording and animal selection as well as researchers and experts from agricultural faculties have contributed the most to the transfer of knowledge to practice during the last decades. Most of the counselling is connected to technology followed by ecology. New ecological criteria require different knowledge and a diverse approach to agricultural production. Milk quotas and negotiated rights to premiums have contributed to changes in agricultural production, which requires the help of all expert services in agriculture and above all of extension service. Knowledge transfer, performed trough lectures, courses, circles, seminars and workshops, have proved to be the most effective method of education meant for agricultural population. Such instructions are provided mostly during winter. There is a lot of individual counselling in office and in field-work. In 2003, the extension service held 4,750 group education meetings, performed about 200,000 individual consultations and prepared more than 1,400 investments, technological and adjustment plans. Agricultural experts provide regular information for farmers using mass media on local and national level. Last year the extension service published over 2,000 expert articles and 27,000 notices. Expert recommendations are avaliable on telephone responder. Furthermore, farmers are educated by organized exhibitions, demonstrations of production and skills, experimental crop growing, organization of expert excursions and various tasting and promotional presentations.

Keywords: transfer of knowledge to practice, extension service, milk recording, education

Introduction

The development of extension service in Slovenia dates back even before the independence in 1991. It functioned within the Cooperative Union of Slovenia and agricultural cooperatives. During the turbulent years of state formation the extension service would have certainly ceased. Hence in 1990 Minister of Agriculture, Forestry and Food decided that the extension service operated under the Ministry of Agriculture, Forestry and Food and remained there until the establishment of Chamber of Agriculture and Forestry of Slovenia (CAFS), it became part of CAFS in 2000 (Klopcic & Osterc, 2003).

Before the formation of the state of Slovenia the extension service personnel transferred to farmers primarily the knowledge about technology that contributed to higher production and thus to higher income and standard of living (MAFF, 2001). In the last decade consumers became aware of better agricultural production, methods and especially of environmental problems caused by intensive industrial way of farming. Consumers are interested in quality as well as healthy and save food. New agricultural methods of production have already been introduced. New requirements have also appeared by the acceptance of Slovenia to the EU respecting EU standards on one hand and the new Slovenian legislation concerning all branches of agriculture on the other. In order to follow the new requirements, the extension service needs to transfer the new knowledge to farmers. The aim of the present article is to show the ways and the extent of transfer of knowledge from public extension service to farmers. The performers of expert tasks also play an important role in the system of knowledge transfer.

Current state of Slovenian agriculture

In the last two decades the number of all kinds of farms in Slovenia decreased at a high rate. Table 1 shows that between the years 1981 and 2000 many farms diminished (over 40 %). Full time farms were halved. According to certain estimations the number is even lower than that shown in Table 1. Today full time farms are market oriented and specialised. Thus they need a close cooperation with the extension service. Part-time farms, which are smaller, do not need so much expert knowledge; they need different knowledge and changed but effective working methods. The preservation of smaller farms would require production of certain agricultural products on the farms, which would contribute to higher added value of market products and consequently to higher income. Concerning rich and diverse nutrition heritage our possibilities in the EU cannot be neglected. Therefore these farms need new knowledge and, subsequently, the extension service personnel need further education.

Table 1. Farms in Slovenia according to socio-economic types, per year.

Type of farm	1981	1991	2000
	No.	No.	No.
Full time	27,986	23,813	14,873
Part-time	54,077	55,797	29,722
Supplementary	53,794	21,529	33,202
Old	13,048	10,812	8,538
Total	148,886	111,951	86,335

Source: SURS and BF, 1982, 1992, 2001

In Slovenia sustainable ways of agricultural production have already been introduced. In the last years organic food production became popular. Over 4 % of agricultural areas where fodder is produced, where 11,000 cattle, 20,000 small ruminants, a few pigs and horses are reared, are under control of organic food production. The organic animal, which is an integrated way of agricultural production, has been extended to include one fifth of all cattle and sheep at present. The breeders, on the other hand, need profound knowledge that was not required in the traditional way of farming. Their needs represent new challenges to the extension service.

Forms of education for extension service personnel
(transfer of knowledge to extension service personnel)

The officers at the Chamber of Agriculture and Forestry of Slovenia (CAFS) are aware that extension service personnel need new knowledge every day. And the personnel themselves are aware of this fact. Therefore a higher or university level of education is required for extension service personnel. Furthermore, they should attend a 100-hour course of special education that is provided by Biotechnical Faculty (Figure 1). After the course a licence to work in extension service is obtained. The extension service personnel should attend organised seminars, symposia, demonstrations and excursions and other forms of self-instruction and education programmes. The above mentioned forms are provided by faculties or by the Chamber and are well attended (Table 2). The excursions are always organised to foreign countries with more developed agricultural production than in Slovenian.

Table 2. Further education of extension service personnel in the year 2003.

Participants in diverse forms of education, No.	2,018
Participants of expert excursions and visits, No.	827

Source: Report of CAFS

Figure 1. Workshop for experts.

Forms of education for farmers - Transfer of knowledge to farmers

The knowledge is transferred to farmers by group education in the form of seminars, courses, trainings, education given to those farmers that do not have adequate qualifications, by publishing expert articles and notices and by individual counselling in the office or on the farm (Table 3).

Data in Table 3 are very illustrative. Lectures, courses or seminars were attended by 22 participants on average. Extension service personnel published a lot of expert articles in the local newspapers, each of them 7 per year on average. They prepared a lot of meetings for radio broadcasts and notices for telephone responders. Nevertheless, personal recommendations in offices and on the farms prevailed.

Table 3. Forms of transfer of knowledge to farmers in the year 2003.

Transfer	Performances, No.	Farmers, No.	Participants, articles, recommendations, notices, No.	Certificates, No.
Lectures, courses, seminars, etc.	4,749		103,485	
Vocational qualification		1,626		678
Expert articles			2,061	
Notices			27,226	
Recommendations in the office		101,092	164,159	
Recommendations on the farms		25,702	55,500	

Source: Report of CAFS

Figure 2. Education for farmers.

The best way of knowledge transfer is to show practical examples. The extension service is aware of this fact. Therefore the personnel organize exhibitions, demonstrations, visits and appraisals (Table 4). All mentioned performances are very well attended.

Table 4. Performances and participants.

Performance	No.	Participants, No.
Exhibitions, various	393	33,155
Visits, various	2,473	23,449
Appraisal of products	1,313	15,626

Source: Report of CAFS

A lot of subsidies are paid by means of Slovenian Agro-Environment Programme (SAEP) measures that were accepted in 2001 by the Slovenian government. Farmers that take part in the SAEP measure need some expert instructions, thus extension service personnel pay a lot of attention to further education. The control and handing of certificates for integrated and organic production also take a lot of work (Table 5). Farmers are interested in SAEP measures therefore a lot of farmers take part in such educational programmes.

Table 5. Educational programme for SAEP and handed certificates in the year 2003.

Farms in SAEP measures, No.	17,342
Participants in educational programmes, No. of handed certificates	20,622
Certificates for integrated fruit production, No.	793
Certificates for integrated wine production, No.	1,261
Certificates for integrated gardening, No.	265
Certificates for organic agricultural production, No.	632

Source: Report of CAFS

Figure 3. Simmental cows on the pasture waiting for ...

Figure 4. Zootechnical Department of Biotechnical Faculty as carrier of education for experts.

Conclusion

In Slovenia the number of farms, especially full-time farms, is rapidly decreasing. Farming on part-time and full-time farms requires profound knowledge on agricultural technology and also new knowledge that especially part-time farms need for supplementary activities. Despite many opportunities for self-instruction most of new knowledge is still transferred to farmers by extension service. Thus the extension service personnel need further education in the programmes of lifelong learning. Well-trained experts can transfer knowledge to farmers. Numerous forms of education and of knowledge transfer provided by extension service show that Slovenian extension service personnel are aware of the importance of their work.

References

Report of CAFS, 2004. Chamber of Agriculture and Forestry (CAFS) – Extension Service of Slovenia. Report of CAFS for year 2003. Ljubljana, Slovenia, 51pp.

MAFF, 2001. Slovenski kmetijsko okoljski program. Ministry of Agriculture, Forestry and Food, Ljubljana, Slovenia, 71pp.

Klopcic, M. & J. Osterc, 2003. Extension work in dairy and beef husbandry in Slovenia. 54th EAAP Meeting, Roma, Italy, 12pp.

Keyword index

Author index

Printed in the United States
by Baker & Taylor Publisher Services